Botanical Keys to Florida's Trees, Shrubs, and Woody Vines

A Guide to Field Identification

Gil Nelson

Pineapple Press, Inc.
Sarasota, Florida

TO BRENDA

Pineapple Press, Inc.
P.O. Box 3889
Sarasota, Florida 34230
www.pineapplepress.com

Library of Congress Cataloging in Publication Data
Nelson, Gil, 1949-
Botanical keys to Florida's trees, shrubs, and woody vines : a guide to field identification / Gil Nelson. -- 1st ed.
    p. cm.
Includes bibliographical references and index.
ISBN 978-1-56164-499-5 (pb : alk. paper)
1. Woody plants--Florida--Identification. I. Title.
QK154.N433 2011
582.1609759--dc23
                          2011029601

First Edition

Design by Shé Hicks
Printed in the U.S.A.

Contents

Acknowledgments

I am sincerely grateful to a number of previous authors and workers, many of whose published works are listed in the References. I am especially indebted to Drs. Richard Wunderlin and Bruce Hansen for their pioneering work in developing the first complete flora of Florida, Dr. Robert K. Godfrey for his exquisitely detailed descriptions of north Florida's trees and shrubs, Dr. Andre Clewell for his guide to the vascular plants of the panhandle, Drs. Donovan and Helen Correll for their complete treatment of the Bahama flora, Dr. Roy Woodbury and colleagues for their excellent work on the trees of Puerto Rico and the Virgin Islands, Dr. P. B. Tomlinson for his keys to the native trees of tropical Florida, and Dr. Alan Weakley for his keys to the plants of the temperate Southeast.

I am also deeply grateful to several regional herbaria, many of which make images of their collections available for study online. The most important of these herbaria for my work include those from Fairchild Tropical Botanic Garden, Florida State University, Tall Timbers Research Station and Land Conservancy, University of Florida, and University of South Florida. I thank Dr. Austin Mast, professor of botany and Director of the Robert K. Godfrey Herbarium at Florida State, for extending numerous considerations, including the opportunity to coordinate the Deep South Plant Specimen Imaging project and complete access to herbarium resources; Dr. Kevin Robertson for my appointment as Beadel Fellow in botany and unrestricted access to the herbarium at Tall Timbers Research Station; and to Lynka Woodbury for hosting me at the Fairchild Tropical Garden during her tenure as herbarium director.

I also offer thanks to Wilson Baker, Keith Bradley, Roger Hammer, Angus K. Gholson Jr., Ron Lance, Dick Wunderlin, and Pat Howell for responding to numerous queries and/or valuable time in the field; and to Marvin for allowing use of his drawings of morphological features from *Trees of Florida*.

I extend deep and continuing appreciation to my wife, Brenda, for her steadfast support and encouragement of my work, my obsessions, and my passions.

The above help notwithstanding, I accept full responsibility for the content of this volume, including any errors, omissions, or shortcomings.

Introduction

This book contains a complete set of dichotomous keys to Florida's trees, shrubs, and woody vines. It may be used as a standalone volume or as a companion to the books *Trees of Florida* and *Shrubs and Woody Vines of Florida*. Included are 968 species, encompassing 508 trees, 628 shrubs, and 132 woody vines. Since some species are counted in more than one category, the total number of species is smaller than the sum of the individual components.

The keys are presented as a series of numbered couplets, each offering a choice between contrasting sets of opposing identification characters. Selecting one or the other petition in each couplet leads to a subsidiary key, another couplet, a family or genus, or a particular species. The couplets are based as far as possible on easily observable features.

The keys are constructed primarily for field use and are not designed to elucidate phylogenetic arrangement or evolutionary relationships. Every effort has been made to rely on features easily seen in living plants, with emphasis on features that can be readily contrasted by observation or measurement and that are available for observation for much of the year. Nevertheless, some species are inherently similar and distinguished only by flowers, fruit, or difficult-to-describe features. This is especially true at the family level. In such cases, it is difficult to avoid relying on somewhat obscure differences or difficult-to-see morphological features.

Arrangement and Format

There are essentially two sets of keys. The first—the Master Key to Major Groups of Woody Plants—leads to one of 11 subsidiary keys, or in a few instances directly to a particular plant family. The subsidiary keys lead to families, genera, or occasionally directly to species. The master keys are followed by family keys, each leading to identification of the species within a particular family. In many cases, species within particular genera are grouped together within the family keys. However, no effort has been made to construct keys that lead to or isolate genera.

Families are presented alphabetically by Latin name, without regard to phylogeny or higher categories of classification. This arrangement makes it easier to find one's way. Whenever possible, I have followed the most recent family classification as presented by the Angiosperm Phylogeny Group (http://www.mobot.org/mobot/research/apweb/), a classification that relies on the latest phylogenetic consensus. Other nomenclatural references include *Guide to the Vascular Plants of Florida*, Third Edition, *Flora of North America* (http://fna.huh.harvard.edu/), and *Plant Systematics: A Phylogenetic Approach*, the latter largely an expression of APG classification. Relying on APG classification has necessitated the inclusion of some genera and species within non-traditional or potentially unfamiliar families. Readers are encouraged to consult the index for genus names that appear to be absent from the keys.

Each family key is followed by the list of taxa referenced in its key, again in alphabetical order by Latinized name. Each species account includes the Latinized name followed by the author(s) of the name, one or more widely accepted common names, statements regarding habitat and distribution in Florida, the native range of the plant if not native to Florida, whether the species is endangered or threatened, and, where appropriate, its invasive status as reported by the Florida Exotic Pest Plant Council. Non-native species are denoted by a leading asterisk (*). Non-native species that are also listed as invasive by the Florida Exotic Pest Plant Council are further denoted by the symbol ◆. Endangered or threatened species are noted with a leading !. Endemic species (those species whose entire range is confined to Florida) are denoted with a leading •. Species lacking a prepended designation are considered native to Florida.

Two indexes are provided, one to genus, the other to common names. The extensive table of contents and alphabetical arrangement of the book serve as indices to families.

Glossary of Common Botanical Terms

Achene. A dry, single-seeded, indehiscent fruit.

Actinomorphic. Said of a flower that is radially symmetrical, or having the symmetry of a wheel with spokes.

Acuminate. Tapering to a pointed apex, the sides of the taper concave; more or less pinched to a point.

Acute. Tapering to a sharp point forming an angle of less than 90 degrees.

Adventitious. Said of abbreviated limbs that grow along the trunk or other main branches.

Alternate leaves. Leaves that arise singly from the stem rather than in pairs or whorls.

Anaerobic. Generally used to refer to soils that lack free oxygen.

Anther. The pollen-bearing portion of a stamen.

Anthesis. The time at which a flower is open and in full bloom.

Anthocarp. A small, single-seeded fruit enclosed by fused petals and sepals or a receptacle.

Apex. The distal tip of a structure; often used when describing a leaf.

Apices. Plural of apex.

Appressed. Pressed flat, or nearly so, against another structure.

Arborescent. Taking the form of a tree; expressing a single trunk, upright habit, and sufficient height.

Arcuate. Curved in the shape of an arc; often used to describe leaf veins that curve along the leaf margin.

Aril. Pulpy appendage to a seed, as in the red covering of a seed in the Magnoliaceae.

Ascending. Said of a plant or appendage that is curving or pointing upwards at an angle of less than 90 degrees.

Axil. The angle formed where two plant parts are joined; commonly used in reference to the angle between leaf and stem.

Axillary. In or arising from an axil.

Basal. At the base.

Biennial. A plant whose life cycle is completed in two years.

Bipinnate. Doubly pinnate.

Bloom. A waxy, whitish covering sometimes found on leaves, stems, or other plant parts, i.e., glaucous.

Bract. A typically (but not always) reduced, leaflike structure that is normally situated at the base of a flower.

Calyptra. A hood, cap, or lid, especially relating to the flowers of some species in the Myrtaceae.

Calyx. The sepals of a flower referred to collectively.

Cambium. Layer of soft tissue between the bark and the wood which adds width to a trunk or branch.

Campanulate. Said of a flower that has fused petals and is bell-like in shape.

Catkin. A spikelike inflorescence bearing small unisexual flowers; often dangling from a branch.

Caudate. Having a long, taillike appendage.

Cauline. Said of leaves that occur along a stem rather than basally.

Ciliate. Having hairs along the margins of a leaf or other structure.

Clasping. Said of a leaf whose base partially encircles the stem.

Complete. A flower that contains all basic parts, including sepals, petals, stamens, and pistil.

Compound leaf. A leaf divided into smaller leaflets along a common axis or emanating from a central point.

Conspecific. Said of two taxa belonging to the same species.

Cordate. Said of a structure (usually a leaf) that is heart-shaped at the base.

Corolla. A whorl of flower parts separating the sepals from the stamens in a perfect flower; sometimes but not always showy. The petals of a flower when referred to collectively.

Corymb. A branched, flat-topped inflorescence in which the lower branches are progressively longer than the upper, the outer flowers opening first.

Crenate. Said of leaf margins with rounded teeth.

Cuneate. Wedge-shaped.

Cuspidate. With a short, sharp apex.

Cyme. A flower structure in which the distal or apical flowers bloom first.

Cymose. Arranged in a cyme.

dbh. Diameter at Breast Height (1.4 m in the U. S.); used to express the standard diameter of a tree.

Deciduous. Trees or shrubs that shed their leaves each year and remain leafless for most of the winter.

Dehiscence. The time at which a fruit or stamen sheds its contents.

Dentate. Having teeth that are perpendicular to rather than angled from the supporting margin.

Digitate. Palmately compound, with several structures (e.g. leaf segments) arising from a common point.

Dioecious. Said of plants that have unisexual male and female flowers produced on separate plants; hence, plants either male or female.

Distal. Generally used to denote the point that is farthest from the point of attachment, as in the distal end (apex) of a leaf.

Drupe. Fleshy fruit in which the inner wall is hardened, forming one or more pits (stones), each enclosing a single seed.

Ecotone. The transition zone between adjacent plant communities.

Endemic. Occurring only in a particular locality.

Entire. Said of a margin that is smooth rather than toothed.

Epiphytic. Said of a plant that grows on the bark of another plant but does not obtain food from and is not parasitic on its host; such plants are called epiphytes.

Equisetum-leaved. Having leaves reduced in size and reminiscent of members of the genus *Equisetum*.

Evergreen. Trees and shrubs that remain green in winter, the old leaves being replaced immediately with new ones; hence the plant is never devoid of leaves.

Fascicle. Bundled or tightly bound cluster, such as a fascicle of pine needles.

Fastigiate. Narrow and upright in growth form.

Filiform. Threadlike and slender; usually rounded in cross section.

Foliaceous. Having leaflike foliage.

Follicle. A podlike fruit developing from a single carpel and splitting along one seam at maturity.

Gall. A tumorous outgrowth of a stem, leaf, or other structure, often resulting from damage inflicted by insect brooding and food structures.

Glaucous. Covered with a whitish bloom that can be removed by rubbing.

Glabrous. Lacking pubescence; thus smooth, not hairy.

Globose. Rounded like a globe.

Habit. Overall appearance or growth form of a plant.

Halophyte. A plant that grows in salty or alkaline conditions, such as plants of the salt-marsh.

Hastate. In the form of an arrowhead, ordinarily used to describe leaves with basal lobes projecting at nearly right angles from the leaf axis.

Haustoria. Parasitic outgrowths of the roots through which food is absorbed from a host plant.

Head. A crowded cluster of flowers at the tip of a single flower stalk; common in the aster family (Asteraceae).

Herbaceous. Not woody, and dying to the ground each year.

Hypanthium. A tubular or cuplike structure of the flower, the rim of which gives rise to the sepals, petals, and stamens.

Hypocotyl. The part of a germinating seed that emerges from the seed following the radicle and lifts the growing tip, including the embryonic leaves, above the ground or beyond the seedcoat.

Indehiscent. Said of a fruit that does not split at maturity.

Inflorescence. Used variously to refer to the flowering portion of a plant, to the type of flower arrangement, or to a flower cluster.

Infructescence. The inflorescence when fruits have replaced the flowers; a collection of fruits.

Involucre. A cuplike series of bracts subtending a flower or fruit.

Irregular. Said of a flower with bilateral symmetry, meaning that it can be cut through the center in only one way to form equal halves; like some flowers of the bean (Fabaceae) family.

Lanceolate. Lance-shaped, wider at the base and tapering toward the apex, entire structure appearing narrow.

Leaflet. An individual blade on a compound leaf.

Lenticel. A raised pore in the stem of a woody plant, often marked with a corky outgrowth and usually borne on young bark.

Lenticellate. Having lenticels.

Lepidote. Covered with small scales; often used to describe a leaf surface.

Locular. Said of an ovary or fruit that has compartments or cavities (locules).

Monoecious. Said of species having unisexual flowers with both male and female flowers produced on the same plant.

Mucro. A short, sharp point extending beyond the margin.

Mucronate. Bearing a mucro.

Oblanceolate. Reverse of lanceolate; widest portion near the apex rather than the base.

Obovate. Opposite of ovate; two-dimensional shape in the outline of an egg but with the widest portion toward the apex.

Opposite leaves. With leaves arising from the stem in pairs opposite one another.

Orbicular. Circular in outline.

Ovate. A two-dimensional shape in the outline of an egg; widest toward the base.

Palmate. Radiating from a single point like the fingers of a hand, as in palmately compound leaves or palmate venation.

Panicle. A loosely branched, compound inflorescence with stalked flowers.

Papilionaceous. Said of a flower of the subfamily Papilionoideae of the bean family (Fabaceae), characterized by an upright standard petal (inserted outside of two lateral petals in bud), two lateral petals, and a keel composed of two, often fused, lower petals (the "typical" pea flower).

Pedicel. The stalk of a flower.

Peduncle. Stalk of an inflorescence, or the stalk of a flower when the flower is borne singly.

Peltate. Stalked from the center rather than the edge; like an umbrella.

Perennial. Plants that persist through more than two growing seasons.

Perianth. The collective name for the sepals and petals of a flower.

Pericarp. The wall of a fruit.

Perfect. Said of a flower with both stamens and pistils.

Petal. A unit of the corolla.

Petiole. The stalk of a leaf.

Phyllodes. An expanded petiole that resembles a leaf in form and function; produced by some species in lieu of true leaves.

Pinna. A leaflet or other major division of a pinnately or multiply compound leaf.

Pinnae. Plural of pinna.

Pinnate leaf. A compound leaf with leaflets along opposite sides of a central stalk.

Pistil. The ovary, style, and stigma collectively; female portion of a flower.

Pome. A fleshy fruit in which the ovary of the flower becomes surrounded by an enlarged floral tube, such as an apple or pear.

Prickle. A sharp-tipped outgrowth of the epidermis; contrast with spine and thorn.

Prop root. Said of the aerial roots of a red mangrove that extend to the ground from the lower trunk and branches and provide stability to the tree.

Protandrous. Said of a flower in which anthers release pollen prior to the stigmas become receptive, effectively preventing or reducing the possibility of self-pollination.

Proximal. Generally used to denote the point closest to the point of attachment; e.g. the basal end of a leaf.

Pubescent. Covered with soft hairs.

Punctae. Depressed dots in the surface, often in reference to the leaf surface, but also to other parts; usually require magnification to see.

Punctate. With depressed dots (punctae) in the surface; often requires magnification to discern.

Pyrene. The small, bony structure at the center of a drupe containing the seed.

Raceme. An inflorescence with a single axis in which the basal flowers open first.

Rachis. The main axis of a compound leaf.

Receptacle. The basal part of a flower, attached to the pedicle and to which the flowers or flower parts are attached.

Reticulate. Forming a network; often used to describe leaf venation.

Revolute. With rolled-under margins; usually said of a leaf.

Rhombic. Having the outline of a rhombus; diamond-shaped in outline.

Riparian. Said of a plant that is typically situated along the banks of a river or other body of water.

Rosette. A radiating cluster; often refers to leaves radiating from the base of a plant.

Rotate. Said of a flower that spreads radially and in one plane.

Ruderal. Said of highly disturbed habitats such as roadsides, vacant lots, and fields.

Rugose. With a roughened, veiny surface; usually refers to the surface of a leaf.

Samara. A winged fruit.

Scabrid. Said of a structure that is rough to the touch due to stiff hairs.

Scandent. Climbing or vinelike.

Schizocarp. A specialized, multiparted fruit that splits into several capsules at maturity; especially well developed in the Euphorbiaceae.

Sepal. Member of the outermost whorl of flower parts (calyx).

Serrate. Said of a margin that is toothed rather than smooth or lobed.

Sessile. Without a stalk.

Sinuate. With strongly wavy margins.

Spadix. A spike of flowers partially embedded in a swollen, fleshy axis.

Spatulate. Having the outline of a spatula.

Specific epithet. The species designation, or second word, in a binomial scientific name.

Spine. A sharp-pointed, often needlelike structure derived from a leaf or stipule; contrast with prickle and thorn.

Spinescent. Having spines.

Stamen. The anther and filament collectively; the male reproductive organ of a flower.

Staminode. A sterile stamen; usually with diminutive anthers.

Standard petal. The uppermost petal in a papilionaceous bean flower; also called the banner petal.

Steephead. A steep-sided ravine located along a natural gradient and created by the seepage of ground water near the base of the ravine.

Stellate. Star-shaped; often refers to leaf hairs that radiate in a starlike pattern.

Stigma. The part of the flower that receives pollen.

Stilt root. Same as prop root.

Stipe. The stalk that supports a fruit or pistil.

Stipule. A leaflike appendage at the base of a leaf or petiole.

Stone. See pyrene.

Style. The part of the flower, usually somewhat elongated, that connects the stigma with the ovary.

Syconium (syconia, plural). A fleshy, hollow receptacle containing multiple flowers and fruit on the inside wall; best represented by the fig.

Syncarp. An aggregate or multiple fruit derived from numerous ovaries of a single flower.

Tepal. Said of petals and sepals when not easily distinguishable from one another.

Terete. Rounded in cross section.

Terminal. Referring to an appendage that is situated at the apex of a structure.

Thorn. A modified, sharp-tipped, usually relatively short stem; compare with prickle and spine.

Thryse. A compound panicle inflorescence in which an indeterminate central axis is subtended by several laterally disposed determinate cymes.

Tomentose. Densely covered with short, matted, or wooly hairs.

Trifoliolate. A compound leaf with three leaflets.

Tripinnate. A compound leaf with a central axis, one to several secondary axes, with the leaflets attached to the secondary axes.

Truncate. Having a flat or squared-off apex or base.

Umbel. A compound inflorescence in which the several flower stalks arise from a common point.

Utricle. A small, indehiscent, single-seeded fruit with a thin, more or less inflated covering.

Whorl. A circular arrangement of like parts, as leaves or flowers, around a point on an axis.

Wing petal. The lateral petal in a papilionaceous bean flower.

Zygomorphic. Bilaterally symmetrical or asymmetrical (see irregular flower above).

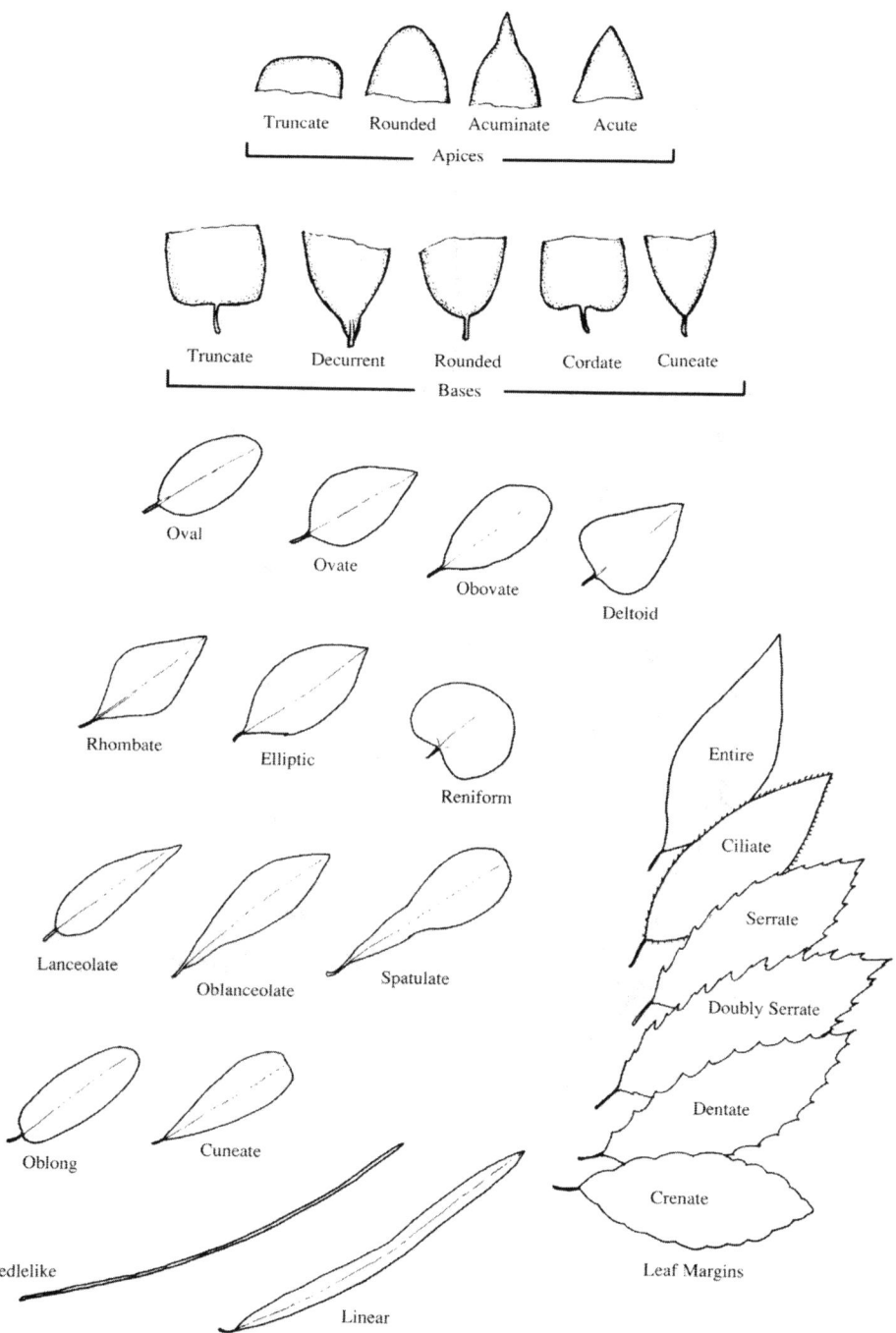

Truncate   Rounded   Acuminate   Acute

Apices

Truncate   Decurrent   Rounded   Cordate   Cuneate

Bases

Oval

Ovate

Obovate

Deltoid

Rhombate

Elliptic

Reniform

Entire

Ciliate

Serrate

Lanceolate

Oblanceolate   Spatulate

Doubly Serrate

Dentate

Oblong

Cuneate

Crenate

Needlelike

Linear

Leaf Margins

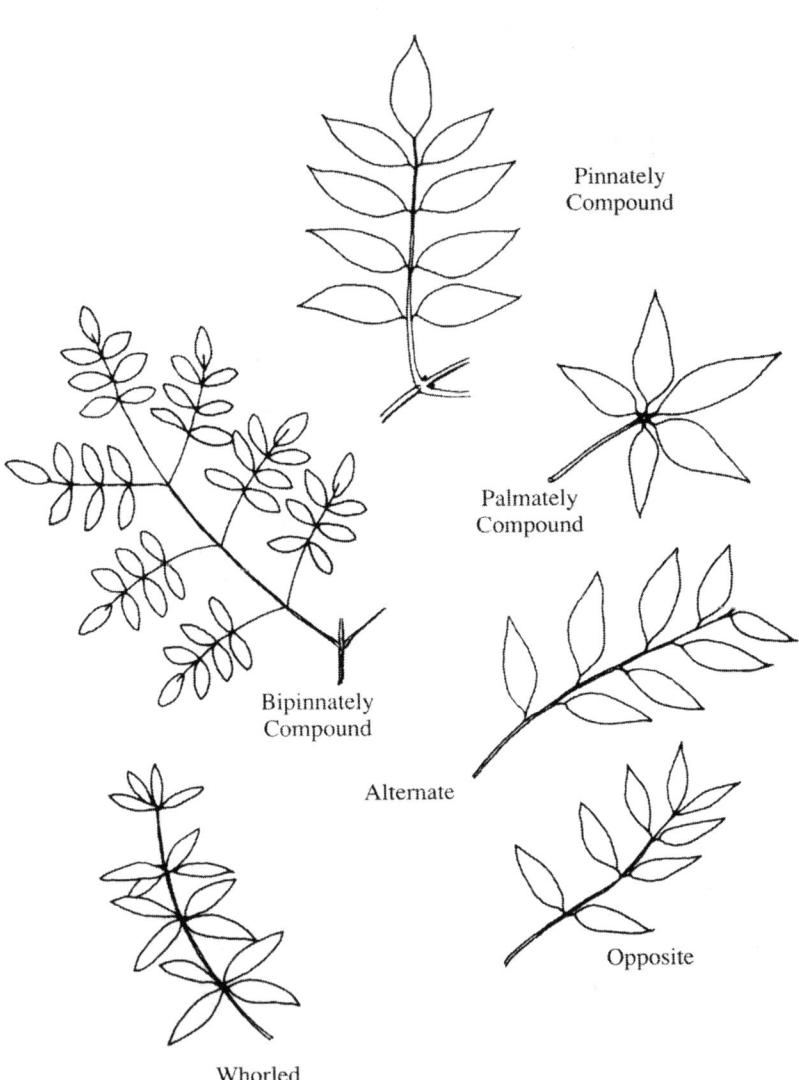

Pinnately
Compound

Palmately
Compound

Bipinnately
Compound

Alternate

Opposite

Whorled

References

Books

Clewell, Andre F. 1985. *Guide to the Vascular Plants of the Florida Panhandle*. Tallahassee: University Presses of Florida, Florida State University Press.

Correll, D. S. and H. B. Correll. 1982. *Flora of the Bahama Archipelago*. A. R. G. Gantner Verlag.

Godfrey, Robert K. 1988. *Trees, Shrubs, and Woody Vines of Northern Florida and Adjacent Georgia and Alabama*. Athens: University of Georgia Press.

Judd, W. S., C. S. Campbell, E. A. Kellog, P. F. Stevens, and M. J. Donoghue. 2008. *Plant Systematics: A Phylogenetic Approach, Third Edition*. Sunderland, MA: Sinauer Associates, Inc.

Lance, R. 2004. *Woody Plants of the Southeastern United States: A Winter Guide*. Athens: University of Georgia Press.

Long, R. W., and Olga Lakela. 1976. *A Flora of Tropical Florida*. Miami: Banyan Books.

Weakley, Alan S. 2011. *Flora of the Southern and Mid-Atlantic States*, Working Draft of 15 May 2011. Chapel Hill: University of North Carolina Herbarium (http://www.herbarium.unc.edu/FloraArchives/WeakleyFlora_2011-May-nav.pdf).

Wunderlin, R. P., and B. F. Hansen. 2008. *Atlas of Florida Vascular Plants* (http://www.plantatlas.usf.edu/).

Wunderlin, R. P., and Bruce Hansen. 2003. *Guide to the Vascular Plants of Florida, second edition*. Gainesville: University Press of Florida.

Website

http://efloras.org/index.aspx A collection of professionally published keys and descriptions.

Master Key to Major Groups of Woody Plants

1. Leaves needle-, scale-, or palmlike; swordlike with a piercing apex; filiform; parallel-veined and grasslike; or absent
  2. Leaves absent, trunk succulent, plant a cactus..Cactaceae
  2. Leaves present, trunk not succulent, plant not a cactus
    3. Leaves needlelike, scalelike, or filiform...Key 1
    3. Leaves not as above
      4. Leaves parallel-veined, grasslike…………………………...…………Poaceae
      4. Leaves not as above
        5. Leaves palmlike or fernlike...Key 2
        5. Leaves swordlike, whorled at ground level or from the top of a short, thick trunk…………………………………………………………...……....Agavaceae
1. Leaves broad, not as above
  6. Leaves opposite or whorled
    7. Leaves (at least some) whorled, simple, entire..Key 3
    7. Leaves opposite
      8. Leaves compound...Key 4
      8. Leaves simple
        9. Margins entire..Key 5
        9. Margins toothed (at least partially) or lobed
          10. Margins toothed, not lobed..Key 6
          10. Margins lobed (toothed or entire)
            11. At least some sepals large, petallike, petals minute.....................
              …………………………………………………….Hydrangeaceae
            11. Sepals and petals all minute……….......................Sapindaceae (*Acer*)
  6. Leaves alternate
    12. Leaves compound
      13. Leaves pinnately or trifoliolately compound...Key 7
      13. Leaves bipinnately compound..Key 8
    12. Leaves simple
      14. Margins entire...Key 9
      14. Margins toothed (at least partially) or lobed
        15. Margins toothed, not lobed...Key 10
        15. Margins lobed (toothed or entire)..Key 11

Key 1: Leaves needlelike, scalelike, filiform, or, if plant cone-bearing, narrowly linear

1. Leaves needlelike, filiform, or, if plant cone-bearing, very narrowly linear
  2. Leaves needlike and borne in sheathed fascicles of 2–5..................................Pinaceae
  2. Leaves needlelike, filiform or narrowly linear, not sheathed
    3. Leaves alternate
      4. Leaves filiform, needlelike, or narrowly linear, evergreen, fruit a cone or a fleshy covering surrounding a single seed
        5. Leaves to about 1 cm wide...Podocarpaceae
        5. Leaves narrower, not exceeding about 5 mm wide, usually narrower
          6. Leaves borne singly, not overlapping on the stem, all of one type............ ....Taxaceae
          6. Leaves spirally arranged, overlapping and obscuring the stem, of 3 types, some awl-shaped, some needlelike, some more or less ovate, mature trees conical in outline with layered branching.…...…...……….....Araucariaceae
      4. Leaves borne on short shoots, these deciduous and falling with leaves intact.…... ……………………………………………..........Cupressaceae (*Taxodium*)
    3. Leaves opposite, often in fascicle-like clusters at the nodes...............Hypericaceae
1. Leaves scalelike or awl-shaped
  7. Leaves opposite or whorled
    8. Plant a multi-branched shrub, stems slender, angled, wiry, usually lax and drooping; flowers bright red, tubular, 1–2.5 cm long…................................. …………………………………….....…….Plantaginaceae (*Russelia equisetiformis*)
    8. Plant a tree, not with the above combination of characters
      9. Leaves whorled, scalelike, grayish, borne 6–17 per node on needlelike twigs.… ……………………………………………………………...Casuarinaceae
      9. Leaves opposite...Cupressaceae
  7. Leaves alternate...Tamaricaceae

Key 2: Leaves palmlike or fernlike (Palms, Cycads, Coontie)

1. Leaves pinnately compound, produced in a terminal whorl on a subterranean or short, thick vertical trunk…….....………………………......…..............Cycadaceae, Zamiaceae
1. Leaves not as above…………………………………………………………..Arecaceae

Key 3: Leaves (at least some) whorled, simple, entire

1. Leaf stalks oozing milky sap when broken...Apocynaceae
1. Leaf stalks with watery sap
  2. Leaves heart-shaped, cordate, or truncate at base; fruit a long, narrow capsule
    (follicle)..…......Bignoniaceae (*Catalpa bignonioides*)
  2. Leaves not as above
    3. Flowers borne in pendent, globular, densely compact, long-stalked, pincushion-
      like heads 2–4 cm in diameter.................Rubiaceae (*Cephalanthus occidentalis*)
    3. Flowers not as above
      4. Flowers red to reddish-orange outside, yellowish inside, tubular and more or
        less angled or ribbed lengthwise, 1.5–4 cm long, flaring at the apex into 5
        lobes, borne in conspicuous axillary or terminal clusters, leaves broadly elliptic
        and with reddish veins and petioles..........................Rubiaceae (*Hamelia patens*)
      4. Flowers purple, pink, or white; leaves lanceolate or narrowly elliptic, tapering
        to both ends, lateral veins conspicuous, numerous, parallel, terminating at a
        submarginal vein that parallels the leaf margin.......................................
        ....…....Apocynaceae (*Nerium oleander*)

Key 4: Leaves opposite, compound (with 2 or more leaflets)

1. Plant a vine
  2. Fruit an achene..…....................Ranunculaceae
  2. Fruit a capsule or berry
    3. Fruit a capsule..…....…........Bignoniaceae
    3. Fruit a berry..............................….............…......................…....Oleaceae (*Jasminum*)
1. Plant a tree or shrub
  4. Leaves palmate
    5. Leaflets predominately 3 (rarely or occasionally 5)
      6. Leaflets entire
        7. Leaflets stalked
          8. Leaflet stalks of equal length, less than 5 mm long, leaflets narrowly
            elliptic or lanceolate.................….......…........Lamiaceae (*Vitex agnus-castus*)
          8. Stalk of terminal leaflet to about 1 cm long, leaflets ovate....................
            ....…....Rutaceae (*Amyris*)
        7. Leaflets all sessile or with very short stalks of nearly equal length..............
        ....…......................Lamiaceae (*Vitex trifolia*)
      6. Leaflets finely or coarsely toothed, or shallowly lobed
        9. Leaflets finely toothed, stalk of terminal leaflet 1.5–3 cm long.................
        ....…..............Staphyleaceae (*Staphylea trifolia*)

9. Leaflets coarsely toothed or shallowly lobed, leaves resembling those of poison ivy (*Toxicodendron radicans*)..............Sapindaceae (*Acer negundo*)
5. Leaflets more than 3
　10. Leaflets entire...Bignoniaceae (*Tabebuia*)
　10. Leaflets toothed..Sapindaceae (*Aesculus pavia*)
4. Leaves pinnate or bipinnate
　11. Leaves pinnate
　　12. Leaflets 4... Zygophyllaceae (*Guaiacum officinale*)
　　12. Leaflets more than 4
　　　13. Leaflets entire, or if toothed, then teeth few and mostly obscure
　　　　14. Leaflets numbering 9–17; flowers large, showy, orange; fruit an elongated pod (follicle) to about 30 cm long............................
　　　　..Bignoniaceae (*Spathodea campanulata*)
　　　　14. Leaflets predominantly fewer; flowers and fruit not as above
　　　　　15. Leaflets numbering 6–10, entire; flowers small, blue; fruit a capsule to about 1.5 cm long...
　　　　　...Zygophyllaceae (*Guaiacum sanctum*)
　　　　　15. Leaflets numbering 5–9, entire or obscurely toothed with few teeth; flowers tiny, green, borne in fascicles or racemes; fruit a 1–2-seeded samara.......................................Oleaceae (*Fraxinus*)
　　　13. Leaflets toothed..Adoxaceae (*Sambucus nigra*)
　11. Leaves bipinnate.............................Bignoniaceae (*Jacaranda mimosifolia*)

Key 5: Leaves opposite, simple, margins entire

1. Plant a vine
　2. Stems oozing milky sap when broken...Apocynaceae
　2. Stems not oozing milky sap when broken
　　3. Plant a sprawling, vinelike succulent of saline environments.............Bataceae
　　3. Plant not a succulent of saline environments
　　　4. Stems armed with piercing, recurved prickles; plant sprawling, shrublike.........
　　　..Nyctaginaceae
　　　4. Stems not armed
　　　　5. Flowers yellow..Gelsemiaceae
　　　　5. Flowers bluish with a yellow throat or purple
　　　　　6. Flowers purple, leaves sandpapery to the touch..............................
　　　　　...Verbenaceae (*Petrea volubilis*)
　　　　　6. Flowers bluish or purple with a yellow throat, leaves smooth to the touch...
　　　　　...Acanthaceae (*Thunbergia*)
1. Plant a shrub, tree, sprawling vinelike succulent, or woody epiphyte
　7. Plant a woody epiphyte, growing in well-elevated tree branches, or a sprawling

succulent of saline environments

8. Plant a woody epiphyte, growing in well-elevated tree branches..........Viscaceae

8. Plant a sprawling, vinelike succulent of saline environments................Bataceae

7. Plant a terrestrial shrub or tree

9. Leaves predominantly greater than 10 cm long

10. Blade heart-shaped (sometimes also lobed in *Paulownia*)

11. Upper surfaces of leaves densely hairy, fruit a hard, ovoid or ellipsoid capsule to about 5 cm long.............Paulowniaceae (*Paulownia tomentosa*)

11. Upper surfaces of leaves sparsely hairy or glabrous, fruit a narrow, elongated pod splitting along a single seam.....................................
....Bignoniaceae (*Catalpa bignonioides*)

10. Blade not heart-shaped

12. Plants of typically saline environments, often standing in or very near salt water (true mangroves)

13. Leaves 5–12 cm long, grayish-green, lower surfaces with grayish pubescence, upper surfaces often displaying visible salt crystals; fruit a flattened, shiny green pod to about 5 cm long; seeds germinating in soil...Acanthaceae (*Avicennia germinans*)

13. Leaves 4–15 cm long, dark green above, lower surfaces covered with numerous black dots, fruit an egg-shaped capsule, germinating while still hanging on the tree and producing a green radicle.....................
....Rhizophoraceae (*Rhizophora mangle*)

12. Plants not of saline environments

14. Leaves obovate, thick, stiff, cardboardlike, typically notched at the apex; petioles swollen at base, producing a distinctive cavity that encloses and protects newly forming parts, oozing yellow sap when broken...Clusiaceae (*Clusia rosea*)

14. Combination of leaves and petioles not as above

15. Lateral veins finely parallel, arising from the midvein at nearly 90°

16. Lateral veins coalescing near the margin into a continuous submarginal vein that virtually outlines the leaf, apex of leaf rounded.........................Apocynaceae (*Ochrosia elliptica*)

16. Lateral veins not coalescing into a submarginal vein, central vein below yellowish, apex of leaf notched, bark of mature trees pitted...Clusiaceae (*Calophyllum*)

15. Lateral veins not finely parallel, wider apart, arising from the midvein at various angles, usually much less than 90°

17. Flowers lacking either sepals or petals, sometimes both

18. Flowers lacking petals (and sometimes sepals), yellow, borne in tight, fascicle-like clusters; fruit a reddish-brown, wrinkled, elongated drupe to about 1.5 cm long..............
....Oleaceae (*Forestiera acuminata*)

18. Flowers with petals present, sepals lacking, reddish- or purplish-brown, borne in axillary cymes; fruit a rounded,

reddish or purplish drupe to about 1 cm diameter.............
....Santalaceae (*Santalum album*)
17. Flowers with both petals and sepals
    19. Petals white, narrowly straplike (inflorescence appearing
        fringlike), borne in a dangling panicle, the stalk of which
        bears leaflike bracts........Oleaceae (*Chionanthus virginicus*)
    19. Petals and inflorescence not as above
        20. Leaves subtended by stipules or stipular line scars that
            connect the bases of the petioles.................Rubiaceae
        20. Stipules lacking or rudimentary, or at least not
            interpetiolar and not leaving stipular line scars
            21. Most parts of the plant (leaves, flowers, fruit) with
                scattered, conspicuous glandular dots (requires
                magnification); flowers with more than 12 stamens..
                ....Myrtaceae
            21. Plants lacking conspicuous glandular dots, stamens
                fewer
                22. Flower petals 5 or more, flowers sometimes
                    double
                    23. Twigs square in cross section, flowers and
                        fruit borne in elongated, dangling, terminal
                        or axillary spikes...
                        ....Verbenaceae (*Citharexylum spinosum*)
                    23. Twigs round in cross section, inflorescence
                      otherwise
                      24. Leaves with 3–5 longitudinal veins, all
                        arising from the base of the leaf and
                        converging at the apex, connected by
                        numerous distinct or obscure transverse
                        lateral veins...............Melastomataceae
                    24. Leaves not as above
                      25. Inflorescence a dense, flat-topped
                        cluster; fruit a drupe....................
                        ....Adoxaceae (*Viburnum nudum*)
                    25. Inflorescence a loose cluster; fruit a
                      berry..........................Oleaceae
        22. Flower petals 4
             26. Twigs square in cross section, foliage often
                 strongly aromatic........................Lamiaceae
             26. Twigs round in cross section.........Oleaceae
9. Leaves predominantly less than 10 cm long
  27. Leaves needlelike, sessile...Hypericaceae
  27. Leaves not as above
    28. Leaves with glands embedded in the blade, these visible with

magnification when held against transmitted light; appearing as tiny yellowish or amber pockets within the leaf issue................Hypericaceae

28. Leaves not as above

29. Leaves opposite and slightly sub-opposite, some leaves more nearly alternate

30. All leaves less than 5 cm long, leaf tip notched, flowers lacking obvious petals

31. Fruit black....................Rhamnaceae (*Krugiodendron ferreum*)

31. Fruit red..Nyctaginaceae (*Guapira*)

30. Some leaves greater than 5 cm long, leaf tip round or pointed, flowers with frilly, brightly colored petals

32. Plant a non-native tree or large shrub, cultivated, not colony forming, rarely escaped into upland habitats......................
....Lythraceae (*Lagerstroemia indica*)

32. Plant a low-growing native wetland colonial shrub, often forming large colonies in standing water wetlands...............
....Lythraceae (*Decodon*)

29. All leaves opposite

33. Most parts of the plant (leaves, flower, fruit) vested with conspicuous glands or glandular dots, these often containing aromatic compounds (requires magnification)

34. Stamens more than 12..Myrtaceae

34. Stamens fewer than 12.............................Malpighiaceae

33. Plant parts lacking conspicuous glandular dots

35. Leaves with arcuate venation, the ends of the primary lateral veins curving upward toward the leaf apex and paralleling the margins, excreting stringy material when the leaf is creased and transversely pulled part..............................Cornaceae

35. Leaves lacking above combination of characters

36. Tips of some leaves notched

37. Leaves less than 5 cm long

38. Flowers unisexual and borne on separate plants, fruit fleshy, red, berrylike, enclosing an achene.......
....Nyctaginaceae (*Guapira*)

38. Flowers bisexual, fruit a rounded, black drupe.........
....Rhamnaceae

37. At least some leaves to 8 cm long...................................
...................Combretaceae (*Laguncularia racemosa*)

36. Leaf tips not notched

39. Lateral veins largely obscure

40. Branches appearing jointed due to the presence of conspicuous interpetiolar scars..............................
...................................Rubiaceae (*Erithalis fruticosa*)

40. Branches not appearing jointed

41. Often branching in 3s, with 2 lateral branches and a central branch arising at a single node, leaves often clustered near the branch tips, the leaf pairs often subtending a pair of sharp, axillary thorns, fruit a white or greenish-white berry....................Rubiaceae (*Randia aculeata*)

41. Plant not branching in 3s, thorns absent, fruit a rounded, purplish-black drupe...........................
............................Oleaceae (*Ligustrum sinense*)

39. Lateral veins largely conspicuous

42. Flowers with 5 blue or white petals, fruit a rounded, bright yellow drupe...Verbenaceae (*Duranta erecta*)

42. Combination of flowers and fruit not as above

43. Interpetiolar stipules present..........…....Rubiaceae

43. Stipules absent

44. Leaves varying elliptic to oval, dull green above with conspicuously depressed, lighter-colored veins..........................
…............Nyctaginaceae (*Pisonia rotundata*)

44. Leaves not as above….............Acanthaceae

Key 6: Leaves opposite, simple, margins at least partially toothed

1. Leaves opposite, alternate, or whorled, often on the same tree; blades ovate, 6–20 cm long, 5–15 cm wide; sap milky; new twigs and petioles with spreading, transparent hairs; fruit a dense, rounded head of red or orange achenes.................................
….......…..Moraceae (*Broussonetia papyrifera*)

1. Plant lacking the above combination of characters

2. Twigs and younger branches green......................................Celastraceae (*Euonymus*)

2. Twigs and younger branches brown, or at least not green

3. Petals free to base..Celastraceae

3. Petals fused at least at base

4. Ovary inferior, flowers and fruits in terminal or subterminal, flat-topped, compound cymes, petals white…....................Adoxaceae (*Viburnum*)

4. Ovary superior, flowers and fruits in axillary fascicle-like clusters, terminal panicles, or drooping axillary racemes, petals purple, greenish-white, or absent

5. Flowers borne in congested, fascicle-like axillary clusters on wood of the previous season, typically prior to the emergence of new leaves, petals absent…..Oleaceae (*Forestiera*)

5. Flowers not with the above combination of characters
  6. Flowers long-stalked, 1 or several in the leaf axils, stamens 5...............
  ....Veronicaceae
  6. Flowers in drooping axillary racemes, spikes, heads, spikelike or open
    racemes, or terminal cymes or panicles, petals present
    7. Flowers in heads, spikes, or spikelike or open racemes; style
      unbranched, stigmatic surface conspicuous.....................Verbenaceae
    7. Flowers in branched axillary or terminal clusters (cymes); style with
      2 branches, the stigmatic surface not conspicuous...............Lamiaceae

Key 7: Leaves alternate, pinnately or trifoliolately compound (with 2 or more leaflets)

1. Plant a vine..Vitaceae
1. Plan a shrub or tree
  2. Fruit a legume..Fabaceae
  2. Fruit not a legume
    3. Leaves with distinctive, amber-colored glands embedded in the leaf tissue, best
      seen with magnification, with leaf backlit.................................Rutaceae
    3. Leaves lacking embedded, amber-colored glands
      4. Leaves palmately compound..Araliaceae
      4. Leaves pinnately, bipinnately, or trifoliolately compound
        5. Leaves trifoliolately compound (with 3 leaflets)
          6. Leaflets (especially the terminal leaflet) stalked
            7. Leaflet margins entire............................Fabaceae (*Erythrina herbacea*)
            7. Leaflet margins finely toothed.........Phyllanthaceae (*Bischofia javanica*)
          6. Leaflets sessile..Sapindaceae (*Hypelate trifoliata*)
        5. Leaves (at least some) pinnately compound
          8. Blades 15–33 cm long, somewhat fernlike, pinnately compound or simple
            with several narrow (on some plants very narrow), deeply lobed, opposite
            or alternate divisions connected to one another by a narrow band of tissue
            along the midrib; divisions numbering 7–19, varying 3–12 cm long,
            divided into several deeply cut lobes; green above, silvery-white beneath
            (especially on new growth)....................Proteaceae (*Grevillea robusta*)
          8. Leaves lacking the combination of characters above
            9. Leaflets predominantly 4 (sometimes 2 or 6)
              10. Rachis conspicuously winged....Sapindaceae (*Melicoccus bijugatus*)
              10. Rachis not winged........................Sapindaceae (*Exothea paniculata*)
            9. Leaflets predominantly more than 4
              11. Fruit a hard, husk-covered nut..........................Juglandaceae (*Carya*)
              11. Fruit otherwise

12. Leaves mostly evenly compound, terminal leaflet lacking or apparently lacking on some or many leaves
  13. Leaflets of some or many leaves alternate (or both alternate and opposite; but see couplet 29)
    14. Rachis not winged
      15. Leaflets 1–2 cm long, numerous, as many as 40 per blade.......Picramniaceae (*Alvaradoa amorphoides*)
      15. Leaflets fewer, longer
        16. At least some leaves with more than 10 leaflets
          17. Leaflets on some leaves numbering 8–32......
            ....Meliaceae (*Khaya senegalensis*)
          17. Leaflets numbering less than 15
            18. Leaflets numbering 6–13, 5–15 cm long, pliable, lanceolate in outline, both surfaces dark green, margins entire.........
              ....Sapindaceae (*Sapindus marginatus*)
            18. Leaflets numbering 10–14, 4–8 cm long, stiff, elliptic to oval in outline, dark lustrous green above, gray below, margins entire and often conspicuously revolute...
              ....Simaroubaceae (*Simarouba glauca*)
        16. Leaflets predominantly numbering 10 or fewer
          19. Leaflets 8–20 cm long
            20. Apex of leaflets pointed.....................
              ....Sapindaceae (*Harpullia arborea*)
            20. Apex of leaflets rounded or notched........
              Sapindaceae (*Cupaniopsis anacardioides*)
          19. Leaflets 5–10 cm long.................................
            ....Picramniaceae (*Picramnia pentandra*)
    14. Rachis winged............Sapindaceae (*Sapindus saponaria*)
  13. Leaflets usually opposite
    21. Rachis winged
      22. Leaflets 1–3 cm long, margins crenate.....................
        ....Rutaceae (*Zanthoxylum fagara*)
      22. Leaflets 3–8 cm long, margins entire or sparsely toothed.................Anacardiaceae (*Rhus copallinum*)
    21. Rachis not winged
      23. Blades with 15 or more leaflets, plant of northern Florida.....................Juglandaceae (*Juglans nigra*)
      23. Blades typically with 8 or fewer leaflets, plant of southern Florida......Meliaceae (*Swietenia mahagoni*)
12. Leaves odd compound, with an even number of lateral leaflets and usually a single terminal leaflet, or leaves trifoliolate
  24. Leaves trifoliolate...........................…....…….Anacardiaceae

24. Leaves pinnate (at least some of them)
 25. Leaflets often variegated, margins wavy and toothed......
  ....Araliaceae
 25. Leaflets not variegated, margins not as above
  26. Leaflets spinosely toothed, the teeth resembling those
   of American holly (*Ilex opaca*).........Berberidaceae
  26. Leaflets not as above
   27. Leaves 15–90 cm long, with petioles 2–15 cm
    long; leaflets typically opposite or subopposite, to
    about 41 in number, 2–15 cm long, borne on
    short stalks, generally lanceolate in outline with
    variously shaped bases, margins with 1–2 (–5)
    prominently gland-tipped teeth located near the
    base beneath, terminal leaflet often of a different
    shape than the lateral leaflets........................
    ....…............Simaroubaceae (*Ailanthus altissima*)
   27. Leaves lacking the above combination of
    characters
    28. Margins of leaflets toothed (even if obscurely
     so)
     29. Leaflets 2–4 cm long, some leaves with
      leaflets obscurely toothed near the apex,
      leaflets often both opposite and alternate...
      ....Anacardiaceae (*Spondias purpurea*)
     29. Leaflets longer
      30. Leaflets to about 8 cm long, margins
       crenate............Anacardiaceae (*Schinus*
       terebenthifolius)
      30. Many leaflets greater than 8 cm long
       31. Leaflets numbering 5–15, plant of
        southern Florida (essentially
        Lower Keys)…...............................
        …..Sapindaceae (*Cupania glabra*)
       31. Leaflets numbering 9–31, plant of
        northern Florida....................
        ....Anacardiaceae (*Rhus glabra*)
    28. Margins of leaflets entire
     32. Most leaflets less than 10 cm long
      33. Leaflet stalk reddish, short, less than
       5 mm long...
       ....Anacardiaceae (*Toxicodendron*
       vernix)
      33. Leaflet stalk yellowish-green, greater
       than 5 mm long

34. Leaflet stalks averaging 1.3–2 cm long, leaflets more or less egg shaped with a blunt apex...............
...........Anacardiaceae (*Metopium toxiferum*)

34. Leaflet stalks averaging 1 cm long or less, leaflets elliptic, asymmetrical, apex long-pointed...
.Burseraceae (*Bursera simaruba*)

32. Most leaflets greater than 10 cm long......
...................Anacardiaceae (*Sorindeia madagascariensis*)

Key 8: Leaves alternate, bipinnately or more compound

1. Plant a vine...Vitaceae
1. Plant a shrub or tree
  2. Fruit a legume...Fabaceae
  2. Fruit a multistoned drupe, legumelike or disklike capsule, or papery 3-angled capsule
    3. Trunk, petioles, and sometimes the leaf rachis armed with prickles; leaves bipinnate or tripinnate, closely set near the top of the trunk and spreading umbrellalike; inflorescence large, terminal, compound, to about 1.2 m long with numerous whitish flowers; fruit a 5-stoned drupe 5–8 mm in diameter..............
....Araliaceae (*Aralia spinosa*)
    3. Trunk unarmed, plant not with the above combination of characters
      4. Ultimate leaflets 1–2 cm long; fruit a legumelike capsule 14–45 cm long.........
....Moringaceae (*Moringa oleifera*)
      4. Ultimate leaflets greater than 2 cm long; fruit a drupe or berry
        5. Fruit a drupe or berry
          6. Plant a shrub, margins of leaflets entire, fruit a red berry......................
....Beberidaceae (*Nandina*)
          6. Plant a tree, margins of leaflets toothed, fruit a yellowish or greenish yellow multi-stoned drupe......................Meliaceae (*Melia azedarach*)
        5. Fruit a papery, 3-angled capsule to about 4 cm long.......................................
....Sapindaceae (*Koelreuteria elegans*)

Key 9: Leaves alternate, simple, margins entire

1. Plant a shrub or tree
  2. Twigs terminated by a true terminal bud surrounded by a cluster of several lateral buds, leaves 5 ranked, terminal leaves appearing clustered near the tip of the twig, fruit an acorn...Fagaceae (*Quercus*)
  2. Combination of twigs and fruit not as above
    3. Flowers lacking petals and sepals
      4. At least some petioles subtended or surrounded by a sheath resulting from the fusion of the stipules (ocrea), sheath sometimes reduced to a ring of hairs at the leaf node; plant a tree or shrub...Polygonaceae
      4. Base of petiole lacking a sheath (ocrea) as described above
        5. Flowers borne in a slender, usually long-stalked fleshy spike.........Piperaceae
        5. Flowers not borne in a slender fleshy spike
          6. Plants dioecious, male and female flowers on separate plants, male flowers borne in catkins
            7. Leaves vested with resinous glands on both surfaces (requires magnification)...Myricaceae (*Myrica inodora*)
            7. Leaf surfaces lacking resinous glands, stem with conspicuous lenticels...Leitneriaceae (*Leitneria floridana*)
          6. Plants monoecious, male and female flowers on the same plant and often in the same inflorescence............................Euphorbiaceae, Phyllanthaceae
    3. Flowers with petals and/or sepals
      8. Flowers with either petals or sepals (corolla or calyx) but not both, or petals and sepals essentially undifferentiated in an apparently single whorl (tepals)
        9. Flowers lacking or apparently lacking sepals
          10. Calyx replaced by a bristlelike pappus to about 1 cm long at fruiting, flowers borne in a tightly clustered head subtended by an involucre...Asteraceae
          10. Calyx reduced to a low ridge around the inferior ovary..Schoepfiaceae (*Schoepfia*)
        9. Flowers not as described above
          11. Male and/or female flowers borne inside of and completely covered by a fleshy receptacle (syconium)........................Moraceae (*Ficus, Brosimum*)
          11. Flowers not as above
            12. Flowers and fruit (achene) borne in a large, greenish, ball-like cluster 10–14 cm in diameter.........................Moraceae (*Maclura pomifera*)
            12. Flowers and fruit not as described above
              13. Flowers and fruit densely compacted into dry, rounded, conelike heads or elongated, compacted spikes...................Combretaceae
              13. Flowers and fruit not as above
                14. Flowers lacking petals
                  15. Fruit a capsule or several-seeded berry.........................

..............................Phyllanthaceae, Phytolaccaceae
    15. Fruit a 1-seeded drupe.................Putranjivaceae (*Drypetes*)
  14. Flowers with petals or tepals
    16. Ovaries inferior, venation often arcuate, major lateral veins often curving toward the tip of the leaf and paralleling the margin for some distance...
    ....Nyssaceae (*Nyssa*), Cornaceae (*Cornus alternifolia*)
    16. Ovaries superior, venation not arcuate, foliage typically aromatic, often strongly so
      17. Flowers borne in open or congested cymes, small, predominantly pale green, white, or yellowish; leaves varying elliptic, lanceolate, somewhat obovate, or mitten-shaped...Lauraceae
      17. Flowers solitary, red and conspicuous, or yellow and more or less inconspicuous, fruit a starlike cluster of pods (follicles) ....Illiciaceae
8. Flowers with differentiated petals and sepals
  18. Petals (at least 2) fused at base
    19. Flowers zygomorophic
      20. Flowers one-sided, all 5 petals radiating in a semicircle on the same side of the flower..............................Goodenaceae (*Scaevola*)
      20. Flowers not one-sided
        21. Fruit a legume..Fabaceae
        21. Fruit a capsule, berry, or a dry, corky drupe
          22. Fruit a dry, corky drupe.............................Myoporaceae
          22. Fruit a capsule or berry.....................Bignoniaceae
    19. Flowers actinomorphic
      23. Stamens more than 12
        24. Leaves with amber-colored glands embedded in the leaf tissue, seen best when held up to transmitted light....................Rutaceae
        24. Leaves lacking embedded amber-colored glands
          25. Fruit a green, oblong drupe to about 1 cm long, flowers yellow, borne in clusters along the branch with or before the new leaves, stamens conspicuous..
          ....Symplocaceae (*Symplocos tinctoria*)
          25. Fruit a berry, varying 2–5 cm in diameter, flowers greenish-white to creamy yellow, urn-shaped, stamens not conspicuous...Ebenaceae
      23. Stamens 12 or fewer
        26. Ovary inferior or at least partially so in *Styrax*
          27. Fruit dry, indehiscent or irregularly dehiscent, a capsule (or capsule-like)
            28. Corolla white...Styracaceae
            28. Corolla yellow.......................................Onagraceae

27. Fruit a berry..Ericaceae (*Vaccinium*)
26. Ovary superior
    29. Leaves all alternate
        30. Ribs of calyx with conspicuous, stalked, glandular protuberances................................Plumbaginaceae
        30. Ribs of calyx lacking conspicuous, stalked, glandular protuberances
           31. Stamens not attached to the corolla tube, or if attached, then only near the corolla base
                32. Fruit a dry, brownish capsule.................Ericaceae
                32. Fruit a yellowish- or orange-red capsule 6–9 mm diameter, or a rounded, green capsule to about 1.2 cm in diameter................................Pittosporaceae
           31. Stamens attached to the corolla tube
              33. Stamens borne opposite the corolla lobes
                  34. Fruit a drupelike berry...................Sapotaceae
                  34. Fruit a capsule....................……Polemoniaceae
              33. Stamens alternating with the corolla lobes
                35. Leaves with glandular dots (requires magnification)...........................Myrsinaceae
                35. Leaves lacking glandular dots
                  36. Fruit a berry or capsule.............Solanaceae
                  36. Fruit a follicle (pod) or drupe
                      37. Fruit a narrow, elongated follicle (pod)...............Apocynaceae (*Plumeria*)
                      37. Fruit a drupe
                          38. Fruit yellow, green, or blackish at maturity, 2–4 cm long...............................Apocynaceae (*Thevetia*)
                        38. Fruit translucent white or red at maturity, averaging well under 2 cm long
                            39. Fruit translucent white at maturity..................................Apocynaceae (*Vallesia*)
                            39. Fruit red at maturity.............…...........Boraginaceae (*Bourerria*)
    29. Leaves alternate below, whorled above.........Theophrastaceae
18. Petals free at the base, not fused
    40. Stamens more than 12
        41. Flowers white, with conspicuous clusters of 24 or more threadlike stamens, opening mostly at night.................Capparaceae (*Capparis*)
        41. Flowers otherwise
            42. Major leaf veins parallel...Myrtaceae

42. Major leaf veins pinnate or palmate
    43. Fruit a whorled or conelike collection of follicles
        44. Fruit a star-shaped collection of follicles, flowers red or creamy yellow, tepals up to 2 cm long, foliage strongly aromatic...Illiciaceae (*Illicium*)
        44. Fruit a conelike aggregation of follicles, each with 1 or 2 seeds, each of which is covered by a scarlet or pinkish aril, flowers typically creamy white (or very pale yellow), tepals 2.5–60 cm long..
            ....Magnoliaceae (*Magnolia*)
    43. Fruit not a whorled or conelike collection of follicles
        45. Leaves palmately veined, at least at the base
            46. Fruit a leathery, nearly round, 5-parted capsule to about 4 cm in diameter..
                ....Malvaceae (*Thespesia populnea*)
            46. Fruit a reddish many seeded, ellipsoid berry, 1–1.5 cm in diameter...
                ....Muntingiaceae (*Muntingia calabura*)
        45. Leaves pinnately veined
            47. Leaf surfaces with glandular dots
                48. Fruit a red, fleshy berry; stamens fused into a tube....................Canellaceae (*Canella winterana*)
                48. Fruit a capsule......................................Myrtaceae
            47. Leaf surfaces lacking glandular dots
                49. Corolla and calyx parts in 3s.............Annonaceae
                49. Corolla and calyx parts in 5s....Chrysobalanaceae
40. Stamens 12 or fewer
    50. Leaves opposite and alternate; flowers white, red, pink, or purple, petals stalked, borne in large, showy panicles, stamens usually 6......
    ....……...Lythraceae (*Lagerstroemia*)
    50. Leaves all alternate, flowers and inflorescence not as above
        51. Fruit simple, a dry achene to about 4 mm long.......................
        ....…..….................Surianaceae (*Suriana maritima*)
        51. Fruit not as above
            52. Ovary inferior, fruit a fleshy pome..........................Rosaceae
            52. Ovary superior
                53. Fruit a capsule, or a dry capsule-like drupe
                    54. Fruit winged
                        55. Fruit 2- or 3-winged.......Sapindaceae (*Dodonaea*)
                        55. Fruit 2–5-winged.....................................
                        ....Cyrillaceae (*Cliftonia monophylla*)
                    54. Fruit not winged......Cyrillaceae (*Cyrilla racemiflora*)
                53. Fruit a drupe or berrylike drupe
                  56. Fruit 1–3 seeded

57. Plants with sharp, needlelike spines in most or all of the leaf axils..............Ximeniaceae (*Ximenia*)

57. Plants lacking spines

    58. Stamens more than 5, inserted on the rim of the hypanthium (floral cup)...Rosaceae (*Prunus*)

    58. Stamens usually 5 or fewer, not inserted on the rim of the hypanthium...........Celastraceae

56. Fruit more than 3-seeded

    59. Leaves averaging 10–25 cm long, fruit a mango.....................Anacardiaceae (*Mangifera indica*)

    59. Leaves averaging less than 15 cm long

        60. Petals hooded, the hoods enclosing or partially enclosing the stamens, fruit a 3-lobed, capsule-like drupe that splits at maturity...................Rhamnaceae (*Colubrina*)

        60. Petals not hooded, fruit a multiseeded, fleshy, berrylike drupe..............Aquifoliaceae (*Ilex*)

1. Plant a vine

  61. Stem with tendrils

    62. Stems usually thorny..Smilacaceae

    62. Stems not thorny

        63. Flowers lacking petals (or sepals and petals undifferentiated and consisting of tepals), the single whorl of parts sometimes showy and easily mistaken for petals...Polygonaceae

        63. Petals and sepals distinct.....................................Passifloraceae

  61. Stem lacking tendrils

    64. Petals absent, sepals petallike

        65. Sepals enlarged, usually tubular, at least below...............Aristolochiaceae

        65. Sepals small, radially spreading..............................Menispermaceae

    64. Petals present, usually fused at base.............................Convolvulaceae

Key 10: Leaves alternate, simple, margins at least partially toothed

1. Plant a vine...……........Vitaceae
1. Plant a tree or shrub
  2. Flowers lacking petals and sepals
    3. Fruit many-seeded, seeds furnished at base with a conspicuous tuft of long, silky hairs (coma)..….......Salicaceae
    3. Fruit with a single seed
      4. Leaves with resinous, amber-colored dots on 1 or both surfaces........Myricaceae
      4. Leaf surfaces lacking resinous dots
        5. Nut enclosed or subtended by a leafy bract, a thin papery sack, or hardened, more or less upright catkin that resembles a tiny pine cone............Betulaceae
        5. Nut otherwise
          6. Nut held in a scaly, cuplike involucre (acorn), twigs terminated by a closely set cluster of lateral buds surrounding a terminal bud.................…………………………………………………………….......Fagaceae (*Quercus*)
          6. Nut enclosed in a spiny bur, buds otherwise....Fagaceae (*Castanea, Fagus*)
  2. Flowers with petals and/or sepals
    7. Flowers with petals or sepals, but not both
      8. Leaves asymmetrical at base, more tissue on 1 side of the midvein than the other
        9. Fruit an aggregate of achenes, these either enclosed by a juicy, fleshy sac or in a dryish calyx and coalesced into a ball-like cluster
          10. Fruiting structure an aggregate of juicy, fleshy sacs, each containing an achene; a mulberry.............…...Moraceae (*Morus*)
          10. Fruiting structure a ball-like cluster of dryish calyces enclosing red or orange achenes..............………...........................Moraceae (*Broussonetia*)
        9. Fruit not as described above
          11. Lateral leaf veins straight and ending in the marginal teeth, or forked, with forks ending in marginal teeth...................….......................Ulmaceae
          11. Lateral veins, especially the proximal pair strongly curving or ascending, not ending in the marginal teeth..Cannabaceae
      8. Leaves not asymmetrical at base
        12. Fruit a capsule
          13. Capsule 5-parted...Malvaceae
          13. Capsule more or less than 5-parted..................................Euphorbiaceae
        12. Fruit a berry, fleshy or dry drupe, drupe-like, or crabapple-like
          14. Stems, petioles, one or both surfaces of the leaves, flowers, and fruits demonstrably scaly...…………….....Elaeagnaceae
          14. Stems, petioles, leaf surfaces, flowers and fruit not scaly
            15. Fruit a white, single-seeded, ovoid drupe 1–2.5 cm in diameter.......

....Putranjivaceae
    15. Fruit otherwise
       16. Fruit a several-seeded berry.............................Salicaceae
       16. Fruit crabapplelike, extremely poisonous, plant parts inducing
         severe dermatitis in some people......Euphorbiaceae (*Hippomane*)
7. Flowers with petals and sepals
  17. Petals united, at least at base
    18. Fruit a capsule
      19. Capsule winged, either flattened or pear shaped...........................
....Styracaceae (*Halesia*)
      19. Capsule not winged
        20. Capsule less than 1 cm long
          21. Flower petals yellow
            22. Petals 4, linear; at least some leaves nearly as broad as
              long, margins bluntly crenate; plants flowering in fall or
              winter...Hamamelidaceae
            22. Petals 5, broad; leaves lanceolate, margins conspicuously
              toothed; plants flowering year round............Turneraceae
          21. Flowers lacking the above combination of characters
            23. Fruit more or less rounded, 3-parted.....Styracaceae (*Styrax*)
            23. Fruit more or less egg-shaped, hard and woody,
              5-parted...Ericaceae
        20. Capsule greater than 1 cm long or broad...........................Theaceae
    18. Fruit a drupe
      24. Stamens more than 12, leaves only faintly toothed..........Symplocaceae
      24. Stamens fewer than 12, leaves distinctly (if finely) toothed, or bluntly
        and crenately toothed
        25. Leaves gland-dotted, fruit bright red.............................Myrsinaceae
        25. Leaves not gland-dotted, fruit white, yellow, or orange...............
....Boraginaceae
  17. Petals free, not united
    26. Stamens borne on the rim of the floral cup (hypanthium)...........Rosaceae
    26. Stamens not as above
      27. Stamens more than 12
        28. Leaf tissue with embedded, yellowish glands (best seen when leaf
          is held up to light); fruit a citron...............................Rutaceae
        28. Leaf tissue lacking yellowish glands, fruit a fleshy berry............
....Muntingiaceae (*Muntingia*)
      27. Stamens fewer than 12
        29. Petals 5, hooded, wrapping around and more or less enclosing the
          stamens...Rhamnaceae
        29. Petals not as above
          30. Petals linear, yellow, flowering late fall and winter; ovary
            inferior.....................................Hamamelidaceae

30. Petals not as above, flowering at various times, not restricted to fall and winter; ovary superior
    31. Fruit a capsule
        32. Leaf margins finely toothed from base to apex.............
....Iteaceae
        32. Leaf margins finely toothed mostly on the distal half of the blade...Clethraceae
    31. Fruit a drupe
        33. Fruit typically with a single stone................Celastraceae
        33. Fruit typically with 4 or more stones.........Aquifoliaceae

Key 11: Leaves alternate, simple, at least some leaves with margins lobed (toothed or entire)

1. Twigs terminated with a true terminal bud surrounded by a closely set cluster of several lateral buds, fruit an acorn..Fagaceae (*Quercus*)
1. Plant lacking the above combination of characters
  2. Petioles exuding milky or colored sap when broken
    3. Leaves peltate
      4. Lobes narrow, simple, less than 20 cm long; fruit a 3-sided capsule...............
....Euphorbiaceae
      4. Lobes broad, 20–60 cm long, often pinnately divided; fruit large, 15–45 cm long, pear-shaped, fleshy, turning from green to orange with maturity.............
....Caricaceae (*Carica papaya*)
    3. Leaves not peltate
      5. Petioles with conspicuous glands near the point of leaf attachment................
....Euphorbiaceae (*Aleurites*)
      5. Petioles lacking glands as described above
        6. Fruit a dry capsule..Euphorbiaceae (*Jatropha*)
        6. Fruit fleshy, juicy, or a ball-like cluster of enclosed achenes............Moraceae
  2. Petioles not exuding milky or colored sap
    7. Leaves consistently palmately star-shaped with 5–7 lobes, or somewhat squared in appearance and truncate at base and apex, or some leaves mitten-shaped
      8. At least some leaves mitten-shaped.....................Lauraceae (*Sassafras albidum*)
      8. Leaves not mitten-shaped
        9. Leaves consistently palmately star-shaped with 5–7 pointed lobes..............
....Altingiaceae (*Liquidambar styraciflua*)
        9. Leaves squared, truncate at base and apex.......................................
....Magnoliaceae (*Liriodendron tulipifera*)
    7. Leaves neither consistently star-shaped, nor truncate at base and apex, nor mitten-like

10. Leaves deeply divided
    11. Leaves somewhat fernlike, pinnately compound or simple with several narrow (on some plants very narrow), deeply lobed, opposite or alternate divisions connected to one another by a narrow band of tissue along the midrib; divisions numbering 7–19, varying 3–12 cm long, themselves divided into several deeply cut lobes; green above, silvery-white beneath (especially on new growth)....................Proteaceae (*Grevillea robusta*)
    11. Leaves deeply palmately lobed
        12. Leaves peltate.....................................Cecropiaceae (*Cecropia palmata*)
        12. Leaves not peltate...Malvaceae
10. Leaves merely lobed
    13. Plants armed with sharp-pointed thorns
        14. Fruit a berry, the outer skin prickly.........................…...Grossulariaceae
        14. Fruit a pome, the outer skin smooth.....................Rosaceae (*Crataegus*)
    13. Plants unarmed
        15. Leaves with 3 primary lobes and veins, the 3 main veins arising from the same point at the apex of the petiole (palmate), each also producing lateral veins
            16. Margins mostly entire; flowers hibiscuslike, with several overlapping petals, pinwheel style; stamens and pistils coalesced into an erect column..Malvaceae
            16. Margins with 1 or more crenations or coarse teeth between the major lobes; flowers not as described above
                17. Leaves fairly consistent in shape (if not size), margins with coarse teeth or much smaller lobes between the larger ones; fruit an achene, borne in a dense, rounded, ball-like cluster at the head of a long peduncle......Platanaceae (*Platanaus occidentalis*)
                17. Leaves variable in shape and size, some lobed, some not, margins of some leaves undulate or with coarse crenations, some maplelike, others 5-lobed and resembling the leaves of sweetgum; fruit a 2-valved capsule to about 5 mm long.............
                ....Salicaceae (*Populus alba*)
        15. Leaves pinnately lobed and veined...........Solanaceae (*Solanum torvum*)

Family Keys

Acanthaceae (Acanthus Family)

1. Plant a tree, natural component of coastal saltwater habitats..........*Avicennia germinans*
1. Plant a shrub or vine, generally spreading from cultivation
  2. Plant a woody twining vine or scandent, vinelike shrub
    3. Plant a woody twining vine...*Thunbergia grandiflora*
    3. Plant a scandent, vinelike shrub...*Thunbergia erecta*
  2. Plant a shrub
    4. Calyx with about 15 obscure teeth at the apex; plant sometimes scandent and vinelike...*Thunbergia erecta*
    4. Calyx deeply or shallowly 5 lobed; plant a shrub
      5. Flowers with 2 elongated fertile stamens and 2 smaller, non-functional stamens
        6. Corolla blue, more or less funnel shaped, with a narrow tube below, expanding at the apex into 5 overlapping petals; leaves green...*Eranthemum pulchellum*
        6. Corolla purple or crimson, 2-lipped; leaves green or purplish, variegated or mottled with splotches of creamy yellow.....................*Graptophyllum pictum*
      5. Flowers with 4 fertile stamens, often with 2 long and conspicuous, the other 2 shorter and less conspicuous
        7. Corolla 2-lipped, flowers yellow.......................................*Barleria lupulina*
        7. Corolla with a slender tube and 5 nearly equal petals, flowers purple varying yellowish or white..*Asystasia gangetica*

*◆*Asystasia gangetica* (**Linnaeus**) **T. Anderson**. Chinese Violet. Disturbed sites, escaped from cultivation. Pinellas, Hillsborough, Highlands, and Okeechobee counties southward, including the Keys. India. FLEPPC listed (II).
Avicennia germinans (**Linnaeus**) **Linnaeus**. Black Mangrove. Tidal swamps and wet-lands, mangrove fringe. Bay County southward along the west coast, St. Johns County southward along the east coast, including the Keys.
Barleria lupulina* Lindley. Hophead Philippine Violet. Disturbed sites, escaped from cultivation. Southern peninsula, essentially Monroe County and the Keys. Africa.
Eranthemum pulchellum* Andrews. Blue Sage. Disturbed areas, escaped from cultivation. Essentially Pinellas County. India.
Graptophyllum pictum* (Linnaeus**) **Griffith**. Caricatureplant. Hammocks, especially where disturbed. Essentially Volusia County. New Guinea.
Thunbergia erecta* (Bentham**) **T. Anderson**. Bush Clockvine. Disturbed sites, escaped from cultivation. Miami-Dade County. Africa.
Thunbergia grandiflora* Roxburgh. Skyvine, Clockvine, Bengal Trumpet. Disturbed sites, escaped from cultivation. Collected in Marion and Lee counties, perhaps elsewhere.

Adoxaceae (Mochatel Family)

1. Leaves pinnately compound...*Sambucus nigra*
1. Leaves simple
  2. Leaves coarsely toothed
    3. Leaves palmately lobed and veined, usually with three lobes, faintly reminiscent in shape to leaves of red maple (*Acer rubrum*).....................*Viburnum acerifolium*
    3. Leaves pinnately veined, unlobed, the veins parallel and terminating at the tips of marginal teeth...*Viburnum dentatum*
  2. Leaves finely toothed or entire
    4. Leaves entire, often folded upward from the midrib..............…......*Viburnum nudum*
    4. Leaves finely toothed
      5. Upper leaf surface more or less dull green, margins bluntly toothed or tending toward entire, lower leaf surfaces punctate with numerous tiny dots (requires magnification), sometimes pubescent but lacking patches of rusty hairs
        6. Leaves obovate tending toward spatulate, predominately less than 5 cm long, many leaves distinctly widest toward the apex.................*Viburnum obovatum*
        6. Leaves elliptic or lanceolate, predominately well over 5 cm long, widest at or near the middle..…..*Viburnum nudum*
      5. Upper leaf surface lustrous green, margins finely sharply toothed, lower surface not punctate and with patches of rusty pubescence..............*Viburnum rufidulum*

Sambucus nigra **Linnaeus.** Elderberry, Common Elder, American Elder. Moist or wet roadsides and ditches, wet disturbed sites, hammocks, floodplains, swamp borders. Essentially throughout, except the Keys.
Viburnum acerifolium **Linnaeus.** Mapleleaf Viburnum. Dry to moist slopes and bluffs. Panhandle; Gadsden and Liberty counties westward.
Viburnum dentatum **Linnaeus.** Arrowwood, Southern Arrowwood. Bluffs, slope forests, floodplains, swamp margins. Throughout the panhandle, east and south to Hernando and Volusia counties.
Viburnum nudum **Linnaeus.** Possumhaw. Wet woods, marsh edges, shrub bogs, flatwoods. Throughout the panhandle, east to Nassau County, south to Glades County, mostly absent from east coastal counties south of St. Johns County.
Viburnum obovatum **Walter.** Walter's Viburnum, Small-leaf Viburnum. Floodplain forests, moist calcareous woodlands. Nearly throughout, Walton to Broward and Monroe counties, not including the Keys.
Viburnum rufidulum **Rafinesque.** Rusty Haw, Rusty Blackhaw, Southern Blackhaw. Moist slopes, temperate hardwood forests, margins of floodplains. Panhandle and northwestern peninsula, Escambia to Hernando counties.

Agavaceae (Agave Family)

1. Leaf margins entire or with fine, sharp, inconspicuous outgrowths (enations) that are sharp and tearing to the touch, sometimes producing threadlike filaments
    2. Leaf margins of at least some leaves producing threadlike filaments, plants essentially lacking a trunk, leaves appearing basal, inflorescence elevated 1–3 m above the leaves..*Yucca filamentosa*
    2. Leaf margins lacking threadlike filaments, stems erect, the uppermost leaves borne well above ground level, leaf margins often with at least a few sharp enations
        3. Leaves dark green, at least 1/3 of the flowering panicle seated within and obscured by the uppermost leaves..*Yucca aliofolia*
        3. Leaves bluish green, becoming grayish, the base of the inflorescence level with or elevated to as much as 50 cm above the uppermost leaves..............*Yucca gloriosa*
1. Leaves typically spiny or prickly along part of the margin, sometimes conspicuously so
    4. Leaf apex blunt; flowers drooping
        5. Leaves to about 2.5 m long, their margins entire, at least on the distal half; plants essentially without a visible trunk...*Furcraea foetida*
        5. Leaves to about 1 m long, their margins with curved teeth; plants with a trunk to about 5 m tall...*Furcraea selloa*
    4. Leaf apex abruptly narrowed, terminated by a long, sharp spine; flowers erect or recurved
        5. Leaf blades at least 15 cm broad
            6. Leaf margins variably toothed, the largest teeth 5–10 mm long...........................
            ....*Agave americana*
            6. Leaf margins finely toothed or entire, teeth closely spaced when present...........
            ....*Agave neglecta*
        5. Leaf blades usually not exceeding about 12 cm broad
            7. Leaf margins with conspicuous, regularly spaced, dark brown, hooked spinescent teeth
                8. Flowers 6–8 cm long, ovaries 40–48 mm long; marginal teeth 2–3 mm long, spine at apex of leaf 1–2 cm long; trunk very thick due to thick leaf bases, 15–25 cm diameter..*Agave decipiens*
                8. Flowers 6–7 cm long, ovaries 35–40 mm long; marginal teeth 3–6 mm long, spine at apex of leaf mostly 2–3 cm long; trunk thinner..............
                ....*Agave fourcroydes*
            7. Leaf margins entire or with a few small usually irregularly spaced teeth
                9. Leaf margins usually whitish or yellowish.....................*Agave angustifolia*
                9. Leaves usually green throughout
                    10. Plant comparatively small, leaves arching or recurved, usually 50–80 cm long; flowering scape usually not exceeding about 3 m tall; inflorescence dense, flowers 4–6 cm long......................................*Agave desmettiana*
                    10. Plants larger; leaves arching or ascending, 80–160 cm long; flowering scape usually much exceeding 3 m tall; inflorescence open, flowers 5.5–6.5 cm long..*Agave sisalana*

Agave americana **Linnaeus.** American Century Plant. Disturbed sites, escaped from cultivation. Southwestern peninsula, Sarasota to Lee counties. Texas, Mexico.

Agave angustifolia **Haworth.** Century Plant. Disturbed sites, escaped from cultivation. Southern peninsula, Sarasota and Martin counties southward. Costa Rica to Tamaulipas, Mexico.

•*Agave decipiens* **Baker.** False Sisal. Coastal hammocks, shell middens. Essentially southwestern coastal Florida, Hillsborough County south to Miami-Dade County and the Keys; also reported from Martin County.

Agave desmettiana **Jacobi.** Dwarf Century Plant. Open disturbed sites. Southwest coast, Sarasota and Lee counties. Mexico.

Agave fourcroydes **Lemaire.** Henequen. Disturbed coastal strand. Sarasota County. Mexico.

•*Agave neglecta* **Small.** Wild Century Plant. Coastal woodlands, shell middens. Southeast and southwest coast, Sarasota to Lee counties, St. Lucie to Palm Beach counties.

*◆*Agave sisalana* **Perrine.** Sisal Hemp. Coastal strand. Southern peninsula, Hillsborough and Brevard counties southward, more common along the coast. Mexico. FLEPPC listed (II).

*◆*Furcraea foetida* **(Linnaeus) Haworth.** Mauritius Hemp. Disturbed sites, escaped from cultivation. Tropical America. FLEPPC listed (II).

Furcraea selloa **K. Koch.** Wild Sisal. Disturbed sites, escaped from cultivation. Manatee and Miami-Dade counties. Colombia.

Yucca aloifolia **Linnaeus.** Spanish Bayonet, Aloe Yucca. Coastal strand woodlands, grassy islands, sandhills. Sporadically distributed throughout the state.

Yucca filamentosa **Linnaeus.** Adam's Needle. Sandhills, sandhill pinelands, dry slope forests. Throughout the panhandle south to Lee and Broward counties.

!*Yucca gloriosa* **Linnaeus.** Moundlily Yucca. Depressions in inland dunes, margins of calcareous glades. Rare and sporadically distributed across northern Florida; should be expected from Jackson to Nassau counties, south on the west coast to Pinellas County. State endangered.

Altingiaceae (Sweetgum Family)

Liquidambar styraciflua **Linnaeus.** Sweetgum. Moist hammocks, wet bottoms, slope forests, roadsides and highway medians, bluffs, temperate hardwood forests, disturbed woodlands. Throughout the panhandle and northern peninsula, south to Hendry and Martin counties.

Anacardiaceae (Cashew Family)

1. Leaves simple...*Mangifera indica*
1. Leaves compound
  2. Leaves trifoliolate
    3. Fruit copiously and conspicuously red-hairy, plant rare in Florida, restricted to a few stations in the panhandle...*Rhus aromatica*
    3. Fruit glabrous or inconspicuously short-hairy, green at first, becoming more or less straw colored, plants common and widespread
      4. Leaflets with apices acuminate; plants of various habitats, often where soil is wet or moist; petiole usually not longer than the leaflet blade.....................
      ....*Toxicodendron radicans*
      4. Leaflets with apices blunt or rounded; plants of sandhills and dry woodlands; petiole usually longer than the leaflet blades................*Toxicodendron pubescens*
  2. Leaves pinnate
    5. Leaf rachis winged
      6. Fruit lacking hairs, flowers with 10 stamens..................*Schinus terebinthifolius*
      6. Fruit with conspicuous glandular hairs, flowers with 5 or fewer stamens
        7. Leaflet margins entire or nearly so, rachis conspicuously winged; plant a small tree or large shrub, much exceeding 1 m tall at maturity......................
        ....*Rhus copallinum*
        7. Leaflet margins coarsely and conspicuously toothed, rachis not or only slightly winged; plant a low shrub less than 1 m tall...............*Rhus michauxii*
    5. Leaf rachis not winged
      8. Plant a low shrub, usually not exceeding about 1 m tall at maturity...................
      ....*Rhus michauxii*
      8. Plant a tree or large shrub, much exceeding 1 m tall
        9. Leaflets entire or obscurely and bluntly toothed, especially near the apex
          10. Leaflets sessile or very short stalked, if leaflet stalk present, usually not exceeding about 1 mm long..*Spondias purpurea*
          10. Leaflets, at least the lowermost, with stalks exceeding 1 mm long
            11. Many or most leaflet stalks much exceeding 1 cm long; fruit a yellow-orange ellipsoid drupe averaging about 1 cm long...................
            ....*Metopium toxiferum*
            11. Leaflet stalk short, much less than 1 cm long; some leaflets appearing sessile; fruit a grayish white, dry, rounded drupe to about 6 mm in diameter..*Toxicodendron vernix*
        9. Leaflets toothed
          12. Leaflets 1–4 cm wide, usually numbering 9–31; plants of the panhandle...
          ....*Rhus glabra*

          12. Leaflets 4–6 cm wide, usually numbering 5–11, plants of the southern peninsula...*Sorindeia madagascariensis*

Mangifera indica **Linnaeus.** Mango. Disturbed sites, escaped from plantings. Lower peninsula, Brevard and Pinellas counties southward. Asia.

Metopium toxiferum **(Linnaeus) Krug & Urban.** Poisonwood, Florida Poison Tree. Subtropical hammocks, rocky pinelands. Southeastern peninsula, Martin County south, including the Keys.

Rhus aromatica **Aiton.** Fragrant Sumac. Upland oak–hickory hammocks. Rare and scattered in Florida, essentially Escambia and Jackson counties.

Rhus copallinum **Linnaeus.** Winged Sumac, Shining Sumac. Sandhills, woodlands, margins of fields, roadsides, dry uplands. Throughout.

Rhus glabra **Linnaeus.** Smooth Sumac. Woodland margins, upland hammocks, disturbed sites, roadsides. Scattered and irregularly distributed in the panhandle, west of Jefferson County.

!*Rhus michauxii* **Sargent.** False Poison Sumac, Michaux's Sumac. Dry hammocks. Alachua County. State and federally endangered.

*◆*Schinus terebinthifolius* **Raddi.** Brazilian Pepper, Pepper Tree. Disturbed sites, hammocks, pinelands, coastal woodlands. Peninsula, Franklin and Duval counties southward, including the Keys. Tropical America. FLEPPC listed (I).

Sorindeia madagascariensis **de Candolle.** Mtikiza. Disturbed sites. Southernmost peninsula, essentially Miami-Dade County. Madagascar.

Spondias purpurea **Linnaeus.** Purple Mombin, Hog Plum. Disturbed sites and hammocks, escaped from cultivation. Southern peninsula, essentially Collier County. Tropical America.

Toxicodendron pubescens **Miller.** Atlantic Poison Oak, Eastern Poison Oak. Sandhills, dry upland woods and hammocks. Northern peninsula and panhandle, Marion and Clay counties westward.

Toxicodendron radicans **(Linnaeus) Kuntze.** Eastern Poison Ivy. Hammocks, swamps and swamp margins, wet woods, floodplain forests, margins of springs and spring runs. Throughout.

Toxicodendron vernix **(Linnaeus) Kuntze.** Poison Sumac. Swamps, wet woods, swamp margins, moist to wet flatwoods. Panhandle, central ridge of the peninsula, Highlands County northward.

Annonaceae (Custard Apple Family)

1. Fruit a fleshy aggregate of fused berries borne singly on a stout stalk
    2. Fruit surface smooth...*Annona glabra*
    2. Fruit surface with blunt or sharp protrusions
        3. Fruit surface with blunt protrusions..*Annona squamosa*
        3. Fruit surface with sharp-pointed, toothlike protrusions...............*Annona muricata*
1. Fruit an aggregate of unfused berries borne in clusters of 2 or more at a node
    4. Flower petals in 2 dissimilar series, the inner petals with a pouchlike base
        5. Branchlets zigzag (geniculate)
            6. Fruit black, fleshy, averaging 4–7 mm in diameter...............*Polyalthia suberosa*
            6. Fruit brown or tan, 4–10 cm long..*Asimina angustifolia*
        5. Branchlets more or less straight or slightly curved (sometimes zigzag near the tips of branchlets subtending new growth)
            7. Leaves thin, more or less flexible, generally obovate in outline, most leaves widest above the middle, apices of many leaves acuminate
                8. Leaves averaging greater than 15 cm long (a few shorter than 15 cm), flowers long-stalked, pedicels 1 cm long or longer at maturity, flowers at anthesis 2 cm wide or greater...*Asimina triloba*
                8. Leaves averaging less than 15 cm long (a few longer than 15 cm), flower short-stalked, pedicels less than 1 cm long, flowers at anthesis less than 2 cm wide...*Asimina parviflora*
            7. Leaves leathery, more or less stiff to the touch, apices rounded or bluntly pointed, not at all acuminate
                9. Leaves narrow, mature blade averaging 5–7 times longer than broad..............
                ....*Asimina angustifolia*
                9. Leaves elliptical, obovate, or oblong, blade usually not exceeding about 5 times longer than broad
                    10. Outer surfaces of outer flower petals predominantly white or creamy
                        11. Flowers arising after the emergence of new leaves
                            12. New shoots, petiole, veins of lower surface of leaf, and fruit stalks bright red-hairy; flower buds always terminal on new growth...*Asimina obovata*
                            12. New shoots, petiole, lower surface of leaf, and fruit stalk glabrous or sparsely hairy..*Asimina × nashii*
                        11. Flowers arising before or with the emergence of new leaves
                            13. Inner petals with a yellow spot inside....................*Asimina incana*
                            13. Inner petals with a purple spot inside................*Asimina reticulata*
                    10. Outer surfaces of outer flower petals predominantly light maroon, streaked maroon, or pale pink
                        14. Flowers predominantly with 8 tepals, tepals wholly maroon, plants 1–3 m tall...*Asimina tetramera*
                        14. Flowers predominantly with 6 tepals, tepals streaked maroon

15. Plant usually less than 50 cm tall.......................*Asimina pygmaea*
15. Plant usually exceeding 1 m tall..................….....*Asimina × nashii*
4. Flower petals in 2 similar series, the inner petals lacking a pouchlike base
16. Flowers yellow, petals not recurved.............*Deeringothamnus rugelii* var. *rugelii*
16. Flowers white, petals recurved................*Deeringothamnus rugelii* var. *pulchellus*

Annona glabra **Linnaeus.** Pond Apple. Swamps. Southern peninsula, Brevard and Manatee counties southward, including the Keys.

*Annona squamosa **Linnaeus.** Sugar Apple. Disturbed sites, hammocks, escaped from cultivation. Southernmost peninsula, Manatee and Broward counties southward, including the Keys. Tropical America.

*Annona muricata **Linnaeus.** Soursop. Disturbed sites, perhaps escaped and established in a few locations. West Indies and tropical America.

Asimina angustifolia **Rafinesque.** Slimleaf Pawpaw. Flatwoods, sandy pinelands, sandhills, scrub. North-central peninsula to the western panhandle, Lake and Putnam to Walton counties.

Asimina incana **(W. Bartram) Exell.** Woolley Pawpaw, Polecat Bush. Sandhills and flatwoods. Central and northeast peninsula; ostensibly Gadsden County.

Asimina manasota **DeLaney.** Manasota Pawpaw. Sandhills. Manatee, Sarasota, and Hardee counties. [Recently described; not included in the keys. Distinguished by small flowers with yellow or maroon outer petals, often with revolute margins, maroon inner petals, and drooping, channeled, linear leaves.]

•*Asimina obovata* **(Willdenow) Nash.** Bigflower Pawpaw. Scrub. Cental peninsula, Clay County south to Glades County.

Asimina parviflora **(Michaux) Dunal.** Smallflower Pawpaw. Coastal swales, wet hammocks, upland woods. Northern 2/3 of the state, south to DeSoto and Highlands counties.

Asimina pygmaea **(W. Bartram) Dunal.** Dwarf Pawpaw. Sandhills, flatwoods. North and north-central peninsula, Manatee and Brevard counties north to Madison and Nassau counties.

•*Asimina reticulata* **Shuttleworth ex Chapman.** Netted Pawpaw. Flatwoods, sandhills. Peninsula, Hamilton to Miami-Dade counties.

•!*Asimina tetramera* **Small.** Fourpetal Pawpaw. Coastal scrub. Martin and Palm Beach counties. State and federally endangered.

Asimina triloba **(Linnaeus) Dunal.** Common Pawpaw. Margins of expansive floodplains, moist woods, mesic hammocks. Uncommon and local in the central panhandle, Liberty and Gadsden to Washinton counties.

Asimina × nashii **Kral.** (*A. angustifolia × A. incana*). Sandhills, longleaf pine—scrub oak woodlands. Northern peninsula, east of the Suwannee River.

•!*Deeringothamnus rugelii* **(B. L. Robinson) Small.** Yellow Squirrel-banana. Moist to wet or poorly drained flatwoods, power line easements. Volusia County. State and federally endangered.

•!*Deeringothamnus rugelii* **(B. L.Robinson) Small** var. *pulchellus* **(Small) D. B.Ward.** White Squirrel-banana. Poorly drained flatwoods, power line easements. Orange, Charlotte, and Lee counties. State and federally endangered.

Polyalthia suberosa* **(Roxburgh) Thwaites. Polyalthia. Hammocks. Miami-Dade County. India.

Apocynaceae (Dogbane Family)

1. Plant a vine
  - 2. Leaves mostly in whorls of 4, fruit a spiny capsule...................*Allamanda cathartica*
  - 2. Leaves mostly opposite, fruit a smooth pod (follicle)
    - 3. Stamens fused into a 5-lobed crown (corona) easily visible within the flower at the base of the funnel-shape tube, fruit borne at nearly right angles from the peduncle at maturity
      - 4. Lobes of the crown deeply divided into 2 parts.............*Cryptostegia grandiflora*
      - 4. Lobes of the crown not divided..........................*Cryptostegia madagascariensis*
    - 3. Stamens not fused into a crown
      - 5. Corolla 1 cm long or less
        - 6. Leaves somewhat thin, flexible, deciduous; plants occurring in floodplain woodlands of larger rivers......................................*Trachelospermum difforme*
        - 6. Leaves leathery, stiff, evergreen; plants escaping cultivation and established only in disturbed areas near plantings.............*Trachelospermum jasminoides*
      - 5. Corolla greater than 2 cm long
        - 7. Flowers yellow...*Pentalinon luteum*
        - 7. Flowers white, pinkish white, or creamy white
          - 8. Corolla narrowly tubular below the spreading petal lobes, flower strongly resembling a pinwheel; leaves often reflexed upward from the central axis, the apex pointed and often curving downward.................*Echites umbellata*
          - 8. Corolla more or less funnel- or bell-shaped; leaves elliptic, mostly flat with a rounded apex..*Rhabdadenia biflora*
1. Plant a shrub or tree
  - 9. Leaves alternate
    - 10. Leaves averaging 8 cm or less in length, fruit a glossy, white, nearly translucent ellipsoidal or pear-shaped drupe...*Vallesia antillana*
    - 10. Leaves averaging longer than 8 cm, fruit a yellow, green, or blackish drupe, or a pair of follicles
      - 11. Leaves to 7 cm wide, usually widest toward the apex, apex notched; lateral veins conspicuous, parallel, and regularly spaced, especially when viewed from beneath; petiole 1–4 cm long; fruit a pair of follicles....*Plumeria obtusa*
      - 11. Leaves narrowly linear, not exceeding about 1.5 cm wide; lateral veins obscure; petiole 3–5 mm long; fruit a yellow, green, or blackish drupe...........
      ....*Thevetia peruviana*
  - 9. Leaves opposite or whorled
    - 12. Plant bearing branched, sharp-pointed thorns piercing to the touch......................
    ....*Carissa macrocarpa*
    - 12. Plant lacking thorns
      - 13. Plant a low-growing herbaceous, slightly woody, or woody shrub or vine less than 1 m tall or long
        - 14. Leaves opposite, flowers yellow or creamy yellow, plants to about 1 m

tall..*Angadenia berteroi*
  14. Leaves whorled, flowers white, plants to 2 m tall.....*Rauvolfia tetraphylla*
 13. Plant a shrub or tree, typically much taller than 2 m at maturity
   15. Leaves opposite, flowers always white
    16. Flowers less than 2 cm long.............................*Tabernaemontana alba*
    16. Flowers 2 cm long or greater.................*Tabernaemontana divaricata*
   15. Leaves whorled, flowers varying white, purple, pink, or yellow
    17. Leaves narrowly lanceolate with numerous, closely spaced, parallel
     lateral veins, diverging at nearly right angles from the midrib, and
     positioned about 1 mm apart......................................*Nerium oleander*
    17. Leaves broadly lanceolate, elliptic, or obovate, lateral veins
     somewhat more widely spaced
      18. Leaves oblanceolate, widest near the apex; fruit bright red at
       maturity...*Ochrosia elliptica*
      18. Leaves lanceolate or elliptic, widest near or slightly below the
       middle
        19. Leaves mostly 4 or fewer per node, corolla less than 1 cm
         long...*Alstonia macrophylla*
        19. Leaves mostly 5 or more per node, corolla greater than 1 cm
         long,..*Alstonia scholaris*

Allamanda cathartica **Linnaeus.** Golden Trumpet, Brownbud Allamanda. Disturbed sites, escaped from cultivation. South-central peninsula, Lake and Volusia counties, south to Broward County. Tropical America.

◆Alstonia macrophylla **Wallich ex G. Don.** Deviltree. Disturbed sites, escaped from cultivation. Miami-Dade County. Southeast Asia, including Malaysia, China, Indonesia, Philippines, Thailand, Vietnam. FLEPPC listed (II).

Alstonia scholaris **(Linnaeus) R. Brown.** Dita, White Cheesewood. Disturbed sites. Palm Beach and Broward counties. India to Indonesia, tropical Australia, Africa.

!*Angadenia berteroi* **(Alph. de Candolle) Miers.** Pineland Golden Trumpet. Pine Rocklands. Miami-Dade County and the Keys. State threatened.

Carissa macrocarpa **(Ecklon) A. de Candolle.** Natal Plum. Disturbed sites, escaped from cultivation. Peninsula, mostly along the coast, Volusia and Sarasota counties southward. Southern Africa.

Cryptostegia grandiflora **R. Brown.** Palay Rubbervine. Disturbed hammocks, escaped from cultivation. Keys. Africa.

◆Cryptostegia madagascariensis **Bojer ex Decaisne.** Madagascar Rubbervine. Disturbed hammocks, escaped from cultivation. Southern peninsula, Manatee and Martin counties southward, including the Keys. Madagascar. FLEPPC listed (II).

Echites umbellatus **Jacquin.** Devil's Potato, Rubbervine. Coastal pinelands. Eastern seaboard, Brevard County southward, including the Keys.

Nerium oleander **Linnaeus.** Oleander. Disturbed sites, escaped from cultivation. Widely planted, established in scattered sites from Franklin and Duval counties southward, including the Keys. Mediterranean.

Ochrosia elliptica **Labillardiére.** Elliptic Yellowwood. Disturbed sites, escaped from cultivation. Escambia and Broward counties and the Keys. New Caledonia, Australia.

Pentalinon luteum **(Linnaeus) B. F. Hansen & Wunderlin.** Wild Alamanda, Hammock Viperstail. Mangrove wetlands, coastal hammocks. Southern peninsula, Lee and St. Lucie counties southward, including the Keys.

Plumeria obtusa **Linnaeus.** Frangipani. Disturbed sites, escaped from cultivation. Widely planted, naturalized in the Florida Keys. West Indies.

Rauvolfia tetraphylla **Linnaeus.** Be-still Tree. Disturbed sites, escaped from cultivation. Palm Beach County. Tropical America.

Rhabdadenia biflora **(Jacquin) Müller Argoviensis.** Rubbervine, Mangrove Rubbervine, Mangrovevine. Mangrove wetlands, coastal hammocks. Charlotte and Brevard counties southward, including the Keys.

Tabernaemontana alba **Miller.** White Milkwood. Disturbed pinelands. Miami-Dade County. Central America.

Tabernaemontana divaricata **(Linnaeus) R. Brown ex Roemer & Schultes.** Cape Jasmine, Pinwheel Flower. Disturbed sites, escaped from cultivation. Osceola County. India.

Thevetia peruviana **(Persoon) K. Schumann.** Luckynut. Disturbed sites, escaped from cultivation. Established in widely scattered locations in the southern peninsula from Brevard and Manatee counties southward. Central America, Mexico.

Trachelospermum difforme **(Walter) A. Gray.** Climbing Dogbane. Floodplain forests, margins of sloughs and backwaters. Panhandle, east to Marion County and south to Hernando County.

Trachelospermum jasminoides **(Lindley) Lemaire.** Confederate Jasmine. Disturbed sites, escaped from cultivation. Naturalized at widely scattered locations statewide. China.

!*Vallesia antillana* **Woodson.** Tearshrub. Subtropical hammocks. Southern peninsula, Miami-Dade and Monroe counties, including the Keys. State endangered.

Aquifoliaceae (Holly Family)

1. Leaves stiff, leathery, evergreen
    2. Leaves averaging less than 3 cm long
        3. Leaf margins crenate or bluntly toothed throughout..........................*Ilex vomitoria*
        3. Leaf margins entire, or with a few tiny teeth....................................*Ilex myrtifolia*
    2. Leaves averaging greater than 3 cm long
        4. Leaf margins crenate distally, with 2–4 blunt, notchlike teeth per side, these usually near the apex and always confined to the distal half of the blade................
        ....*Ilex glabra*
        4. Leaf margin entire or more consistently toothed
            5. Marginal teeth or apex tipped with a sharp, piercing spine greater than 1 mm long

      6. Leaf blade dark green, greater than 3 cm wide, the margins flat or undulate, slightly if at all revolute...*Ilex opaca* var. *opaca*

      6. Leaf blade light green or yellow-green, less than 3 cm wide, the margins conspicuously revolute..*Ilex opaca* var. *arenicola*

    5. Margins entire, or if toothed, neither the teeth nor apex armed with a rigid, piercing spine

      7. Margins toothed

        8. Marginal teeth small, spreading, typically widely but regularly spaced along the margin from apex to base; leaves with minute black punctae on the lower surfaces; fruit black...*Ilex coriacea*

        8. Marginal teeth small, angling toward the apex, typically irregularly spaced along the margin; leaves lacking minute black punctations; fruit red or yellow..*Ilex cassine*

      7. Margins entire

        9. Inflorescence stalked, fruit red or yellow...............................*Ilex cassine*

        9. Inflorescence not stalked, fruit black or purplish black...........*Ilex krugiana*

1. Leaves pliable, thin, deciduous, sometimes borne crowded at the tips of short shoots and appearing opposite or whorled

    10. Leaf blades predominately oblanceolate varying to spatulate or obovate, widest toward the tip and tapering to a more or less narrow base

      11. Flower and fruit stalks greater than 1 cm long..................................*Ilex longipes*

      11. Flower and fruit stalks less than 1 cm long

        12. Larger leaves averaging greater than 1.5 cm wide, longer than 5 cm; fruit averaging greater than 5 mm diameter....................*Ilex decidua* var. *decidua*

        12. Larger leaves averaging less than 1.5 cm wide, less than 5 cm long; fruit averaging 5 mm in diameter or less........................*Ilex decidua* var. *curtissii*

    10. Leaf blades oblong, elliptic, oval, or ovate, predominately widest near the middle

      13. Plants of river bottoms, seepages, bogs, swamps, and other wet woodlands

        14. Leaf margins inconspicuously toothed with mostly appressed teeth, venation of lower leaf surface distinctly reticulate, blade soft green or yellowish green; flowers typically with 4 sepals and 4 petals; fruit dull phosphorescent red, to about 10 mm in diameter.................*Ilex amelanchier*

        14. Leaf margins conspicuously toothed, venation of lower leaf surface if reticulate, not distinctly so, blade dark green; flowers typically with 5–8 sepals and 5–8 petals; fruit bright red, to about 7 mm diameter.................... ....*Ilex verticillata*

      13. Plants of dry, well-drained, nearly xeric uplands.......................*Ilex ambigua*

***Ilex ambigua* (Michaux) Torrey.** Carolina Holly, Sand Holly. Sandhills, scrub, dry upland woods, hammocks, sometimes in association with limestone. Throughout the panhandle and northern peninsula, south to Lee and Martin counties.

!*Ilex amelanchier* M. A. Curtis ex Chapman. Sarvis Holly, Serviceberry Holly. Swamps and floodplain woodlands. Panhandle, Liberty County westward. State threatened.

***Ilex cassine* Linnaeus.** Dahoon. Swamps, wet hammocks, coastal wetlands, roadside ditches, pond margins. Throughout, except the Keys.

Ilex coriacea **(Pursh) Chapman.** Large Sweet Gallberry. Wet and moist flatwoods, swamps and swamp margins, bogs and bog margins. Panhandle and northern peninsula, south to Polk County.

Ilex decidua **Walter.** Possumhaw Holly, Deciduous Holly. Floodplain forests of large and small rivers, upland hammocks, sometimes where limestone is present.

Ilex decidua **Walter var.** *curtissii* **Fernald.** Curtis's Holly. Floodplains, swamps, sandy wetlands, usually in soils overlying limestone. Northern peninsula, mostly in association with the Suwannee River.

Ilex glabra **(Linnaeus) A. Gray.** Inkberry, Gallberry. Flatwoods, bogs, coastal wetlands. Throughout, except the Keys.

!*Ilex krugiana* **Loesener.** Tawnyberry Holly, Krug's Holly. Rockland hammocks, pine rocklands. Miami-Dade County. State threatened.

Ilex longipes **Chapman ex Trelease.** Long-stalked Holly, Georgia Holly. Mesic or dry well-drained hammocks. Sporadically and sparsely distributed. Panhandle, Leon County westward.

Ilex myrtifolia **Walter.** Myrtle-leaved Holly, Myrtle Dahoon. Bogs, savannas, margins of flatwoods depressions, swamp margins, coastal wetlands. Panhandle and northern peninsula, south to Alachua and Flagler counties, disjunct to Orange County.

Ilex opaca **Aiton.** American Holly. Rich woods, hammocks, mesic woodlands, slope forests. Panhandle, south to Lake and Hernando counties, disjunct to Palm Beach and Lee counties.

•*Ilex opaca* **Aiton var.** *arenicola* **(Ashe) Ashe.** Scrub Holly. Scrub. Mostly along the central ridge, Baker County south to Glades County.

Ilex verticillata **(Linnaeus) A. Gray.** Common Winterberry. Swamps, wet floodplains. Panhandle, Wakulla County westward.

Ilex vomitoria **Aiton.** Yaupon. Coastal dunes, swamp margins, wet woods, mesic hammocks, flatwoods. Panhandle and northern peninsula, south to Brevard, Highlands, and Sarasota counties.

Ilex × *attenuata* **Ashe.** [NOTE: This taxon and the next are hybrids of *I. opaca* and *I. cassine*; they are not keyed due to insufficient defining characters.]

Ilex × *attenuata* **'East Palatka' Ashe.**

Araliaceae (Ginseng Family)

1. Leaves simple
  2. Plant a vine; leaves cordate at base; blade not deeply divided into several segments... ....*Hedera helix*
  2. Plant a large shrub; leaves peltate; blade deeply divided into several segments.......... ....*Tetrapanax papyriferus*
1. Leaves compound
  3. Leaves palmately compound
    4. Leaves trifoliolate; plant a more or less scandent and vinelike shrub; branches and leafstalks with prickles...*Eleutherococcus trifoliatus*
    4. Leaves with many more than three leaflets; plant a shrub or tree; branches and leafstalks lacking prickles
      5. Leaves with 7–9 leaflets; flowers greenish yellow..............*Schefflera arboricola*
      5. Leaves with 9–15 leaflets; flowers red............................*Schefflera actinophylla*
  3. Leaves pinnately, bipinnately, or more compound
    6. Leaves typically bipinnate; trunk and leaf axis armed with coarse prickles.............. ....*Aralia spinosa*
    6. Leaves typically pinnate; trunk and leaf axis lacking prickles...*Polyscias guilfoylei*

Aralia spinosa **Linnaeus.** Devil's Walking Stick. Dry and moist hammocks, upland and lowland woods, bog margins. Panhandle and northern peninsula, Okaloosa to Hillsborough, Polk, and Osceola counties.

*****Eleutherococcus trifoliatus* (Linnaeus) S.Y.Hu .** Disturbed sites, escaped from cultivation. Alachua County. Asia.

*****Hedera helix* Linnaeus.** English Ivy. Upland and lowland woods, hammocks, often where escaping and invading from nearby plantings. Panhandle and northern peninsula, westernmost panhandle sporadically eastward to Alachua County. Europe.

*****Polyscias guilfoylei* (Bull ex Cogniaux & Marchal) L. H. Bailey.** Frosted Aralia. Disturbed sites, escaped from cultivation. Keys. Malaysia.

*◆*Schefflera actinophylla* (Endlicher) Harms.** Australian Umbrella Tree, Octopus Tree. Disturbed sites, invading coastal and subtropical hardwood hammocks, sand pine scrub, dunes, cypress strand. Australia. FLEPPC listed (I).

*****Schefflera arboricola* (Hayata) Merrill.** Dwarf Schefflera. Disturbed sites, escaped from cultivation. Southeastern peninsula, Martin to Miami-Dade counties, also naturalized in Volusia and potentially other counties.

*****Tetrapanax papyriferus* (Hooker) K. Koch.** Ricepaper Plant. Disturbed sites, escaped from cultivation and established in scattered locations from at least Walton to Hillsborough, Polk, and Brevard counties, perhaps elsewhere.

Araucariaceae (Araucaria Family)

Araucaria heterophylla (Salisbury) Franco. Norfolk Island Pine. Widely cultivated, rarely naturalized in hammocks and disturbed sites. Southern peninsula. Norfolk Island.

Arecaceae (Palm Family)

1. Plant a shrub, stem belowground or only partially emergent; not treelike
  2. Petioles armed with sharp marginal spines..*Serenoa repens*
  2. Petioles unarmed
    3. Tip of petiole extending as a midrib into the leaf blade (costapalmate)
      4. Tip of petiole extending into the blade as much as 1/4 of the blade length, terminating in a triangular extension; leaf segments lacking threadlike fibers along the margins, plants mostly of moist to wet woods and hammocks of the northern peninsula and panhandle..*Sabal minor*
      4. Tip of petiole extending into the blade for much of its length; margins of leaf segments adorned with conspicuous threadlike fibers; plants of the peninsular scrub...*Sabal etonia*
    3. Tip of petiole terminating at the blade
      5. Leaves palmate, tip of petiole rounded (see *Sabal minor*), trunk often emerging aboveground (often nearly concealed by the fronds) for 30 cm or more and covered with a mass of sharp, needlike spines 20–30 cm long....................
        ....*Rhapidophyllum hystrix*
      5. Leaves pinnate..*Chamaedorea seifrizii*
1. Plant a tree
  6. Leaf blades palmate (or apparently so), with blade segments radiating from a (near-) central point
    7. Petiole with sharp marginal spines, protrusions, or teeth for at least part of its length
      8. Petiole spines not exceeding about 2 mm long; plant ordinarily a shrub with a prostrate trunk, sometimes erect and treelike...............................*Serenoa repens*
      8. Petiole spines longer than 2 mm
        9. Petiole spines less than 5 mm long, fronds to about 1 m wide, leaf segments stiff, plants forming distinctive clumps..........................*Acoelorraphe wrightii*
        9. Petiole spines more than 5 mm long, fronds greater than 1 m wide, leaf segments drooping, plants not forming clumps
          10. Petiole split at base, fruit brown and to about 1 cm long, crown often subtended by a shaggy, drooping ring of old fronds...........................
          ....*Washingtonia robusta*

10. Petiole not split at base, fruit bluish and greater than 1.5 cm long
    11. Mature leaves costapalmate, the petiole extending into the blade........
    ....*Livistona chinensis*
    11. Mature leaves palmate, all segments arising from a central point........
    ....*Livistona rotundifolia*
7. Petiole lacking marginal spines or teeth
    12. Blade appearing palmate, but with petiole extending a short distance into the blade (costapalmate) and tapering to a narrow apex, the segments arising along it; blade deeply V-shaped; segments with threadlike fibers along their margins..*Sabal palmetto*
    12. Blade evidently palmate, with all segments arising at a central point; blade not deeply V-shaped
        13. Petiole split at base, inflorescence as long as or longer than the leaves, fruit white
            14. Lower surface of leaves whitish, fruit with conspicuous pedicels to about 5 mm long...*Leucothrinax morrisii*
            14. Lower surface of leaves pale green, pedicel very short, fruit appearing sessile...*Thrinax radiata*
        13. Petiole not split at base, inflorescence shorter than the leaves, fruit varying purple to black...*Coccothrinax argentata*
6. Leaf blades pinnate or bipinnate, with primary segments arising along either side of a central axis, similar in form to that of a feather
    15. Leaves bipinnate, segment margins jagged and reminiscent in shape to that of a fishtail
        16. Jagged portion of blade segment greater than 1/2 the segment length............
        ....*Caryota mitis*
        16. Jagged portion of blade segment less than 1/2 the segment length..............
        ....*Caryota urens*
    15. Leaves pinnate
        17. Petiole with sharp spines, teeth, or protrusions
            18. Petiole with short, angling protrusions to about 5 mm long; fronds grayish, strongly re-curved, the tips nearly reaching the trunk................
            ....*Butia capitata*
            18. Petiole with sharp spines greater than 5 mm long,
                19. Lowermost leaf segments modified into sharp-tipped, yellowish, needlelike spines to about 9 cm long (at least some greater than 3 cm long); spines usually borne in loosely set pairs on either side of the central axis...*Phoenix reclinata*
                19. Petiole margins saw-toothed, lined with hard, sharp spines that are less than 3 cm long and become broader at the base.........*Elaeis guineensis*
        17. Petiole lacking spines
            20. Trunk, leaf blades, and inflorescence armed with needlelike spines.........
            ....*Acrocomia totai*
            20. Trunk, leaf blades, and inflorescence lacking spines

21. Trunk lacking a distinctive crownshaft
    22. Fruit a coconut with a hard outer covering, greater than 20 cm wide, trunk often leaning and curved......................*Cocos nucifera*
    22. Fruit fleshy, trunk typically straight............*Syagrus romanzoffiana*
21. Trunk with a conspicuous green crownshaft
    23. Leaf segments emanating from the midrib in several directions
        24. Inflorescence borne within the leaves, stalk of the inflorescence greater than or equal to 30 cm long................ ....*Pseudophoenix sargentii*
        24. Inflorescence borne below the leaves, stalk of the inflorescence less than 30 cm long.................*Roystonea regia*
    23. Leaf segments emanating from the axis in two directions
        25. Apex of the leaf segments truncate at an angle and raggedly toothed, trunk usually solitary
            26. Segments pleated, lower surface grayish..................... ....*Ptychosperma elegans*
            26. Segments not pleated, lower surface of leaf green........... ....*Ptychosperma macarthurii*
        25. Apex of the leaf segments bluntly pointed, trunks often clustered...*Dypsis lutescens*

!*Acoelorraphe wrightii* (Grisebach & H. Wendland) H. Wendland ex Beccari. Everglades Palm. Swamps, hammocks. Southern peninsula, Collier, Monroe, and Miami-Dade counties, excluding the Keys. State threatened.
*Acrocomia totai Martius. Gru-gru Palm. Disturbed sites, escaped from cultivation. Brevard County. South America.
*Butia capitata (Martius) Beccari. Wind Palm, Jelly Palm, Pindo Palm. Naturalized in disturbed sites, mostly near plantings. Northern and central peninsula. South America.
*Caryota mitis Loureiro. Burmese Fishtail Palm. Disturbed hammocks, escaped from cultivation. Broward and Miami-Dade counties. India to the Philippine Islands.
*Caryota urens Linnaeus. Fishtail Palm. Disturbed hammocks, escaped from cultivation. Miami-Dade County. India.
*◆Chamaedorea seifrizii Burret. Bamboo Palm. Disturbed hammocks. Broward and Miami-Dade counties and the Keys. Guatemala and Belize. FLEPPC listed (II).
!*Coccothrinax argentata* (Jacquin) L. H. Bailey. Florida Silver Palm. Pine rocklands, coastal hammocks, dune margins. Palm Beach to Miami-Dade and Collier counties, including the Keys. State threatened.
*◆Cocos nucifera Linnaeus. Coconut Palm. Escaped from cultivation, widely naturalized along beaches and in disturbed sites. Widely planted northward to the central peninsula, perhaps naturalized only in the southernmost peninsula. Old World. FLEPPC listed (II).
*Dypsis lutescens (H. Wendland) Beentje & J. Dransf. Areca Palm. Disturbed sites, escaped from cultivation. Martin to Miami-Dade counties. Africa, Madagascar.
*Elaeis guineensis Jacquin. African Oil Palm. Disturbed sites, escaped from cultivation.

Miami-Dade County. Tropical Africa.

!*Leucothrinax morrisii* (H. Wendland) C. E. Lewis & Zona. Brittle Thatch Palm, Key Thatch Palm. Tropical hammocks. Monroe and Miami-Dade counties, including the Keys. State endangered.

*◆*Livistona chinensis* (Jacquin) R. Brown ex Martius. Chinese Fan Palm. Disturbed hammocks, escaped from cultivation. Naturalized in scattered locations from Putnam to Miami-Dade counties. Japan, China. FLEPPC listed (II).

**Livistona rotundifolia* (Lamarck) Martius. Footstool Palm. Disturbed hammocks, escaped from cultivation. Miami-Dade County. Malay Peninsula, Java.

*◆*Phoenix reclinata* Jacquin. Senegal Date Palm. Disturbed sites, hammocks, Brevard and Pinellas counties southward along the coast, including the Keys. Africa, Madagascar. FLEPPC listed (II).

!*Pseudophoenix sargentii* H. Wendland ex Sargent. Sargent's Cherry Palm. Coastal woodlands over limestone, especially on islands. Miami-Dade County and the Keys. State endangered.

*◆*Ptychosperma elegans* (R. Brown) Blume. Solitaire Palm, Alexander Palm. Disturbed hammocks, escaped from cultivation. Broward and Miami-Dade counties. Queensland, Australia. FLEPPC listed (II).

**Ptychosperma macarthurii* (H. Wendland ex J. H. Veitch) H. Wendland ex Hook. f. Macarthur Palm. Disturbed hammocks, escaped from cultivation. Miami-Dade County. Australia, New Guinea, Pacific Islands.

Rhapidophyllum hystrix (Pursh) H. Wendland & Drude ex Drude. Needle Palm. Moist hammocks, low woods, slopes above floodplains. Panhandle and northern peninsula, south to Manatee and Highlands counties.

!*Roystonea regia* (Kunth) O. F. Cook. Florida Royal Palm. Swamps and cypress sloughs; widely cultivated. Southern peninsula, DeSoto and Palm Beach counties southward, excluding the Keys. State endangered.

•*Sabal etonia* Swingle ex Nash. Scrub Palmetto. Scrub. Peninsula, Clay to Miami-Dade counties.

Sabal minor (Jacquin) Persoon. Dwarf Palmetto, Bluestem Palm. Floodplains, low woods, slopes, moist or wet hammocks. Panhandle and northern peninsula, south to Charlotte, Glades, and St. Lucie counties.

Sabal palmetto (Walter) Loddiges ex Schultes & Schultes f. Cabbage Palm, Sabal Palm. Hammocks, margins of salt marshes, swamp and bog margins. Statewide. Florida's State Tree.

Serenoa repens (W. Bartram) Small. Saw Palmetto. Flatwoods, margins of coastal and mesic hammocks. Statewide.

*◆*Syagrus romanzoffiana* (Chamisso) Glassman. Queen Palm. Disturbed hammocks, escaped from cultivation. Southern peninsula, Hillsborough and Martin counties, southward. South America, Brazil to Argentina. FLEPPC listed (II).

!*Thrinax radiata* Loddiges ex Schultes & Schultes f. Florida Thatch Palm. Coastal woodlands, especially where limestone is present. Collier, Monroe, and Miami-Dade counties, including the Keys. State endangered.

*◆*Washingtonia robusta* H. Wendland. Washington Fan Palm. Disturbed sites, escaped

from cultivation. Planted statewide, naturalized in Hillsborough County and the south-eastern peninsula. Mexico. FLEPPC listed (II).

Recognized Hybrids
•*Sabal* × *miamiensis* **Zona** (*S. etonia* × *S. palmetto*).

Aristolochiaceae (Birthwort Family)

1. Leaves, at least some on any plant, obovate, widest toward the apex; base of leaf blade truncate to only slightly cordate...*Aristolochia maxima*
1. Leaves ovate, triangular, or circular in outline, widest at or below the middle; base of leaf blade distinctly cordate
  2. Widest part of the calyx gradually expanding from the tube, less than 5 cm wide.......
  ....*Aristolochia maxima*
  2. Widest part of the calyx abruptly spreading from the tube, greater than 6 cm wide
    3. Petallike sepals forming 3 spreading lobes at the apex of the flower; plants native to floodplain forests of the panhandle................................*Aristolochia tomentosa*
    3. Petallike sepals forming 1 or 2 lobes at the apex of the flower; plants not native, escaped from cultivation, north-central and southern peninsula
      4. Petallike sepals forming 2 lobes at the apex of the flower
        5. Upper lobe more or less circular in outline with ruffled margins, bent downward and more or less pendent................................*Aristolochia labiata*
        5. Upper lobe more or less ovate or spatulate, lacking ruffled margins, more or less erect in orientation..*Aristolochia ringens*
      4. Petallike sepals forming a single, continuous lobe
        6. Leaves densely whitish-hairy beneath..........................*Aristolochia gigantea*
        6. Leaves glabrous
          7. Flowers about 10 cm wide; leaf stalks subtended by auriculate, stipule-like structures (pseudostipules)...*Aristolochia littoralis*
          7. Flowers more than 20 cm wide; leaf stalks not subtended by conspicuous pseudostipules...*Aristolochia grandiflora*

***Aristolochia gigantea Martius & Zuccarini.** Calico Flower, Pelican Flower. Disturbed sites. Potentially escaped from cultivation and natualized in southern Florida. Brazil.

***Aristolochia grandiflora Swartz.** Largeflower Dutchman's-pipe. Disturbed hammocks, escaped from cultivation. Southeast peninsula. Tropical America.

***Aristolochia labiata Willdenow.** Mottled Dutchman's-pipe. Disturbed sites, escaped from cultivation. West-central peninsula, essentially Pasco County. South America.

*♦*Aristolochia littoralis* **Parodi.** Callico Flower, Elegant Dutchman's-pipe. Disturbed sites. Peninsula, from about Alachua and Putnum counties south to Miami-Dade County.

Tropical America. FLEPPC listed (II).

Aristolochia maxima **Jacquin.** Dutchman's-pipe. Disturbed hammocks, escaped from cultivation. Miami-Dade County. Mexico, Central America.

Aristolochia ringens **Vahl.** Gaping Dutchman's-pipe. Disturbed hammocks, and dry disturbed sites, escaped from cultivation. Pinellas and Miami-Dade counties. South America.

!*Aristolochia tomentosa* **Sims.** Wooley Dutchman's-pipe, Pipevine. Floodplain forests, panhandle, Gadsden and Liberty counties westward. State endangered.

Asteraceae or Compositae (Aster or Composite Family)

1. Flower heads having both disk and ray flowers (though ray flowers in *Chyrsoma* sometimes few, small, and inconspicuous)
   2. Ray flowers yellow
      3. Leaves alternate..*Chrysoma pauciflosculosa*
      3. Leaves opposite
         4. Leaves grayish; bracts of the inflorescence ridged and bearing an evident sharp spine 1–3 mm long..*Borrichia frutescens*
         4. Leaves green; bracts of the inflorescence soft and not spine-tipped....................
            ....*Borrichia arborescens*
   2. Ray flowers lavender, lavender-pink, or bluish............*Symphyotrichum carolinianum*
1. Flower heads with disk flowers only
   5. Flowers bisexual, both stamens and pistils present
      6. Flowers lavender-pink..*Garberia heterophylla*
      6. Flowers white, varying pinkish white................................*Koanophyllon villosum*
   5. Flowers unisexual
      7. Plants dioecious, staminate and pistillate flowers borne in separate heads on separate plants
         8. Leaves narrowly linear, 1–3 mm wide.............................*Baccharis angustifolia*
         8. Leaves broader
            9. Leaves entire..*Baccharis dioica*
            9. Leaves toothed
               10. Surfaces of leaf blades with conspicuous pale amber punctations (may require magnification); most of the flower heads distinctly stalked.............
                  ....*Baccharis halimifolia*
               10. Surfaces of leaf blades lacking conspicuous punctations, or if punctations present, obscure; most flower heads sessile.........*Baccharis glomeruliflora*
      7. Plants monoecious, staminate and pistillate flowers borne in a single head
         11. Leaves succulent, all but the lowermost alternate.........................*Iva imbricata*
         11. Leaves not succulent, all but the uppermost opposite.................*Iva frutescens*

Baccharis angustifolia **Michaux**. Saltwater Falsewillow. Brackish marshes, margins of salt marshes, beaches, coastal swales, margins of coastal woodlands. Coastal counties, nearly throughout.

!*Baccharis dioica* **Vahl**. Broombush Falsewillow. Coastal hammocks. Miami-Dade County, perhaps extirpated. State endangered.

Baccharis glomeruliflora **Persoon**. Silverling. Floodplains, wet and moist mixed woodlands. Essentially statewide.

Baccharis halimifolia **Linnaeus**. Saltbush, Groundsel Tree, Sea Myrtle. Coastal woodlands, margins of salt marshes, roadsides and other disturbed sites. Throughout.

Borrichia arborescens **(Linnaeus) de Candolle**. Tree Seaside Oxeye. Beaches, margins of salt marshes. Miami-Dade County and the Keys.

Borrichia frutescens **(Linnaeus) de Candolle**. Bushy Seaside Oxeye. Saltmarshes, tidal flats, beaches, salty coastal swales. Coastal counties, essentially throughout.

Chrysoma pauciflosculosa **(Michaux) Greene**. Bush Goldenrod, Woody Goldenrod. Coastal dunes, interior sands of barrier islands, old inland dunes, sandhills. Panhandle, Wakulla County westward.

!•*Garberia heterophylla* **(W. Bartram) Merrill & F. Harper**. Garberia. Sand pine—oak scrub. Central peninsula, Clay to Highlands counties. State threatened.

Iva frutescens **Linnaeus**. Bigleaf Sumpweed. Margins of salt marshes, coastal sand flats, thinly vegetated coastal forests. Coastal counties, essentially throughout; absent from the Keys.

Iva imbricata **Walter**. Seacoast Marshelder. Coastal dunes and beaches. Coastal counties, nearly throughout, including the Keys.

!*Koanophyllon villosum* **(Swartz) R. M. King & H. Robinson**. Florida Shrub Thoroughwort. Hammocks, pinelands. Miami-Dade County, Keys. State endangered.

Symphyotrichum carolinianum **(Walter) Wunderlin & B. F. Hansen**. Climbing Aster. Margins of spring runs, growing on rotting logs above spring waters, swamps, riverbanks. Leon, Franklin, and Wakulla counties southward throughout the peninsula, absent from the Keys.

Hybrids

Borrichia × *cubana* **Britton & S. F. Blake**. (*B. arborescens* × *B. frutescens*)

Bataceae (Saltwort Family)

Batis maritima **Linnaeus** Saltwort, Turtleweed. Salt flats, margins of tidal margins, mangrove, salt marshes. Along the coasts, Bay and Nassau counties south to the Keys.

Berberidaceae (Barberry Family)

1. Leaves bi- or tripinnately compound, leaflets entire, flowers white..................
....*Nandina domestica*
1. Leaves pinnately compound, leaflets toothed, flowers yellow................*Berberis bealei*

Berberis bealei **Fortune**. Beale's Barbary, Leatherleaf Mahonia. Widely cultivated, rarely naturalized in the central panhandle. China.
*◆*Nandina domestica* **Thunberg**. Nandina, Sacred Bamboo, Heavenly Bamboo. Cultivated, escaped, naturalized, rich woods and hammocks. Western panhandle south to the north-central peninsula. China. FLEPPC listed (I).

Betulaceae (Birch Family)

1. Fruiting structure conelike, 1–4 cm long, resembling tiny pine cones
    2. Fruiting "cones" thickened and "woody," persisting on the plant well after the seeds are released, usually into or beyond the following winter, winter buds stalked, leaves in 3 ranks on the branch, plant a shrub at maturity..............................*Alnus serrulata*
    2. Fruiting "cones" leathery, not thickened and "woody," not persisting beyond the season, usually disintegrating with seed fall and leaving only the axis, winter buds stalked, leaves in 2 ranks on the branch, plant a tree................................*Betula nigra*
1. Fruiting structure not conelike, mostly longer than 4 cm, composed of foliaceous bracts, not at all resembling tiny pine cones
    3. Bark smooth, grayish, not shredding, the trunk fluted, mature leaves glabrous above.
    ....*Carpinus caroliniana*
    3. Bark brown, shredding in thin strips, the trunk not fluted, mature leaves finely hairy above..*Ostrya virginiana*

Alnus serrulata **(Aiton) Willdenow**. Hazel Alder. Margins of large alluvial rivers and streams, wetland edges, pond margins, often near standing water where regularly inundated. Panhandle and northern peninsula, south to about Alachua County.

Betula nigra **Linnaeus**. River Birch. Confined mostly to river and stream banks, often where inundated for part of the time. Panhandle to northwestern peninsula, south to about Levy County.

Carpinus caroliniana **Walter**. American Hornbeam, Bluebeech. Low woods, mesic hammocks, margins of spring runs, floodplain woods. Panhandle to south-central peninsula to Manatee and Hardee counties.

Ostrya virginiana **(Miller) K. Koch**. Eastern Hophornbeam. Moist to wet hammocks, mesic to moderately dry slopes, often at mid-slope or above. Panhandle to middle north-central peninsula, south to about Hernando County.

Bignoniaceae (Trumpet Creeper Family)

1. Plant a climbing vine, leaves compound
    2. Leaves usually with 7 or more leaflets, plant lacking tendrils
        3. Flowers varying red to yellow, plant climbing by aerial rootlets...*Campsis radicans*
        3. Flowers pink, inside of tube with purplish red stripes, plants lacking aerial rootlets..*Podranea ricasoliana*
    2. Leaves with 3 or fewer leaflets, plant with terminal tendrils
        4. Tips of the tendrils 3-forked, the forks hooked and clawlike, flower yellow with orange stripes in the throat..*Macfadyena ungui-cati*
        4. Tips of the tendrils not hooked and clawlike (even if 3-forked), flowers orange, reddish orange, or white externally with a yellow throat
            5. Flowers white externally with a yellow throat, or lavender or purple...*Pithecoctenium crucigerum*
                6. Flowers white or creamy white with a yellow throat; leaf tendrils with adhesive disks...*Pithecoctenium crucigerum*
                6. Flowers purple, lavender, or white; leaf tendrils lacking adhesive disks ....*Cydista aequinoctialis*
            5. Flowers orange or reddish orange
                7. Stamens exserted from the corolla, leaf tendrils lacking adhesive disks at their tips, leaflets in 2s and 3s, ovate...............................*Pyrostegia venusta*
                7. Stamens included within the corolla, leaf tendrils with minute adhesive disks at their tips, leaflets in 2s...............................*Bignonia capreolata*
1. Plant a shrub or tree
    8. Leaves simple
        9. Leaves opposite, fruit an elongated follicle...........................*Catalpa bignonioides*
        9. Leaves alternate or closely clustered at the tips of short spur shoots, fruit a large, hard, egg-shaped, or rounded capsule or berry

10. Leaves distinctly alternate, fruit a hard, ellipsoid capsule 7–12 cm long, 4–8 cm broad..*Amphitecna latifolia*

10. Leaves usually congested in fascicle-like clusters at the tips of spur branches, fruit a hard, rounded berry 15–30 cm long, 10–20 cm broad....*Crescentia cujete*

8. Leaves compound

11. Leaves pinnate, bi-, or tripinnate

12. Leave bi- or tripinnate...

13. Leaves bipinnate, leaflets less than 2 cm long, flowers blue.....................
....*Jacaranda mimosifolia*

13. Leaves bipinnate or tripinnate, leaflets at least 4 cm long, flowers white or pale yellow...*Radermachera sinica*

12. Leaves pinnate

14. Leaflets entire, corolla bell-shaped.......................*Spathodea campanulata*

14. Leaflets toothed, corolla tubular or funnel shaped

15. Flowers orange...*Tecoma capensis*

15. Flowers yellow..*Tecoma stans*

11. Leaves palmately compound

16. Flowers yellow...*Tabebuia aurea*

16. Flowers pink, purple, or white......................................*Tabebuia heterophylla*

Amphitecna latifolia (Miller) A. H. Gentry. Black Calabash. Coastal hammocks. Extreme southeastern peninsula.

Bignonia capreolata Linnaeus. Crossvine. Floodplain forests, hammocks, mesic woods. Panhandle south to central peninsula, to about Hillsborough and Polk counties.

Campsis radicans (Linnaeus) Seemann. Trumpet Creeper. Disturbed sites, moist hammocks, floodplain forests. Panhandle to south-central peninsula, to about Collier County.

Catalpa bignonioides Walter. Southern Catalpa. Floodplain forests, moist or wet woodlands, levees along large alluvial streams, widely planted in residential landscapes. Scattered throughout the panhandle and central peninsula, south to about Hillsborough and Polk counties.

Crescentia cujete Linnaeus. Calabash. Naturalized in hammocks and disturbed sites. Florida Keys. Tropical America.

Cydista aequinoctialis (Linnaeus) Miers. Garlic Vine. Disturbed sites. Naturalized in southeast Florida, mostly Broward County. Tropical America.

Jacaranda mimosifolia D. Don. Jacaranda, Black Poui. Planted for ornament in residential landscapes, naturalized in extreme southeast Florida. Peru.

◆Macfadyena unguis-cati (Linnaeus) A. H. Gentry. Cat's-claw Vine. Scattered in locations throughout the state except the Keys. Invasive. Tropical America. FLEPPC listed (I).

Pithecoctenium crucigerum (Linnaeus) A. H. Gentry. Monkey's-comb. Disturbed sites. Extreme southeast Florida. Tropical America.

Podranea ricasoliana (Tanfani) Sprague. Zimbabwe Creeper, Pink Trumpet Creeper. Disturbed sites, escaped from cultivation. East-central peninsula. South Africa.

Pyrostegia venusta (Ker-Gawler) Miers. Flamevine. Disturbed sites, escaped from cultivation. Scattered across the southern half of the peninsula. South Africa.

Radermachera sinica (Hance) Hemsley. China Doll, Serpent Tree. Disturbed sites, escaped from cultivation. Extreme southern peninsula. China, Taiwan.

***Spathodea campanulata* P. Beauvais**. African Tuliptree. Disturbed hammocks, planted for ornament and naturalized in the southernmost peninsula. Tropical Africa.

***Tabebuia aurea* (Silva Manso) Bentham & Hooker f. ex S. Moore**. Caribbean Trumpet-tree. Disturbed sites, escaped from cultivation. Southeastern peninsula. West Indies, tropical America.

***Tabebuia heterophylla* (de Candolle) Britton**. White Cedar. Disturbed sites, escaped from cultivation. Extreme southeastern peninsula and the Keys. Tropical America.

***Tecoma capensis* (Thunberg) Lindley**. Cape Honeysuckle. Disturbed sites, escaped from cultivation. Central and southern peninsula from Hillsborough and Volusia counties southward, including the Keys. South Africa.

***Tecoma stans* (Linnaeus) Jussieu ex Kunth**. Yellow Elder, Yellow Trumpetbush. Disturbed sites, escaped from cultivation. Southernmost and east-central peninsula. Tropical America.

Boraginaceae (Borage Family)

1. Plant a vine
  2. Leaves 10–15 cm long...*Tournefortia hirsutissima*
  2. Leaves 3–7 cm long...*Tournefortia volubilis*
1. Plant a tree or shrub
  3. Leaves mostly linear, densely silky-hairy.................................*Argusia gnaphalodes*
  3. Leaves wider, not densely silky-hairy
    4. Styles twice divided, resulting in each style with 4 ultimate branches, each branch topped with a single stigma
      5. Flowers orange...*Cordia sebestena*
      5. Flowers white
        6. Flowers borne in spikes or rounded clusters
          7. Flowers borne in rounded clusters...................................*Cordia globosa*
          7. Flowers borne in spikes
            8. Leaves conspicuously toothed.................................*Cordia curassavica*
            8. Leaves entire or minutely and inconspicuously toothed..........................
            ....*Cordia bahamensis*
        6. Flowers borne in branched inflorescences resembling a corymb or cyme.......
        ....*Cordia dichotoma*
    4. Styles not twice divided
      9. Leaves bluntly or sharply toothed
        10. Leaves 1.5–3.5 cm long, margins crenate, upper surfaces white-spotted, plant a shrub..*Carmona microphylla*
        10. Leaves 5–13 cm long, margins serrate, upper surfaces not white-spotted, plant usually a tree..*Ehretia acuminata*

9. Leaves entire
11. Upper surfaces of leaves glabrous..............................*Bourreria succulenta*
11. Upper surfaces of leaves distinctly pubescent
12. Petiole 1–3 mm long, fruit 7–8 mm in diameter.......*Bourreria cassinifolia*
12. Petiole 2–7 mm long, fruit 9–14 mm in diameter...........*Bourreria radula*

!*Argusia gnaphalodes* **(Linnaeus) Heine.** Sea Rosemary, Sea Lavender. Dunes. East coast, Brevard County to the Keys. State endangered.

!*Bourreria cassinifolia* **(A. Richard) Grisebach.** Smooth Strongbark, Little Strongbark, Strongback. Pinelands. Southernmost peninsula and the Keys. State endangered.

!*Bourreria radula* **(Poiret) G. Don.** Rough Strongbark, Rough Strongback. Hammocks. Florida Keys. State endangered.

!*Bourreria succulenta* **Jacquin.** Bahama Strongbark, Bodywood, Strongbark, Strongback. Hammocks. Extreme southeastern peninsula and the Keys. State endangered.

*Carmona microphylla **(Lamarck) G. Don.** Fukien Tea. Disturbed sites, escaped from cultivation. Extreme southeastern peninsula. Asia.

Cordia bahamensis **Urban.** Bahama Manjack. Pine rocklands. Very rare. Miami-Dade County.

*Cordia curassavica **(Jacquin) Roemer & Schultes.** Black Sage. Disturbed sites escaped from cultivation. Extreme southeast Florida, Broward County. Central America.

*Cordia dichotoma **Forst. f.** Fragrant Manjack. Disturbed sites, escaped from cultivation. Southeastern peninsula, Palm Beach to Miami-Dade counties. China.

!*Cordia globosa* **(Jacquin) Kunth.** Curaçao Bush. Hammocks. Southernmost peninsula. State endangered.

*Cordia sebestena **Linnaeus.** Geiger Tree. Coastal hammocks, cultivated. Southernmost peninsula. Treated as native by some observers. West Indies.

*Ehretia acuminata **R. Brown.** Koda. Mesic hammocks, especially where disturbed. Alachua County. China, Japan, Australia.

!*Tournefortia hirsutissima* **Linnaeus.** Chiggery Grapes. Hammocks. Southernmost peninsula. State endangered.

Tournefortia volubilis **Linnaeus.** Twining Soldierbush. Coastal hammocks, shell middens. Peninsula from about Hillsborough and Volusia counties southward, mostly along the coast.

Buddlejaceae (Butterflybush Family)

1. Leaves ovate, broadly elliptic, or sometimes nearly circular in outline, the apex blunt...
....*Buddleja indica*
1. Leaves lanceolate or ovate-lanceolate, the apex long-tapering
   2. Leaves distinctly and conspicuously white tomentose beneath............................
....*Buddleja madagascariensis*
   2. Leaves grayish tomentose beneath..*Buddleja lindleyana*

Buddleja indica **Lamarck.** Indoor Oak. Disturbed scrub, escaped from cultivation. Southeast peninsula. Madagascar.
Buddleja lindleyana **Fortune ex Lindley.** Lindley's Butterflybush. Disturbed sites, escaped from cultivation. Widely planted, potentially naturalized throughout. China.
Buddleja madagascariensis **Lamarck.** Madagascar Butterflybush. Disturbed sites, escaped from cultivation. Madagascar.

Burseraceae (Gumbo-Limbo Family)

Bursera simaruba **(Linnaeus) Sargent.** Gumbo-limbo. Tropical coastal hammocks, shell middens. Southern peninsula, from about Pinellas and Volusia counties southward, mostly along the coast.

Cactaceae (Cactus Family)

!*Pilosocereus polygonus* **(Lemaire) Byles & G. D. Rowley.** Key Tree Cactus. Tropical hammocks. Rare and restricted to the Keys. State and federally endangered.

Calycanthaceae (Sweetshrub Family)

1. Lower leaf surface, twigs, and petioles distinctly pubescent..................................
....*Calycanthus floridus* var. *floridus*
1. Lower leaf surface, twigs, and petioles glabrous or sparsely pubescent with a few
   scattered hairs..*Calycanthus floridus* var. *glaucus*

!*Calycanthus floridus* **Linnaeus**. Sweetshrub, Carolina Allspice. Mesic hammocks and slopes. Panhandle sparingly eastward and scattered to about Alachua County. State endangered.
!*Calycanthus floridus* **Linnaeus var. *glaucus* (Willdenow) Torrey & A. Gray**. Eastern Sweetshrub, Carolina Allspice. Mesic hammocks and slopes. Rare in a few counties of the central and western panhandle. State endangered.

Canellaceae (Wild Cinnamon Family)

!*Cunella winterana* **(Linnaeus) Gaertner.** Wild Cinnamon, Pepper Cinnamon, Cinnamon Bark. Tropical hammocks. Southernmost peninsula and the Florida Keys. State endangered.

Cannabaceae (Hemp Family)

1. Lower surface of leaf distinctly pubescent on tissue, large veins, and veinlets (as seen
   with 10× magnification)
    2. Leaves 1–3 cm long...*Trema lamarckiana*
    2. Leaves 6–19 cm long
        3. Leaves much longer than wide, the apex long tapering; fruit black...........................
        ....*Trema orientalis*
        3. Leaves about twice as long as wide, the apex not long tapering; fruit orange.........
        ....*Trema micrantha*
1. Lower surface of leaf glabrous or sparsely pubescent on the larger veins (as seen with
   10× magnification)
    4. Branches armed with short spines; flowers and fruit in cluster of 3 or more
        5. Upper surface of the leaf scabrous (rough to the touch).........*Celtis ehrenbergiana*
        5. Upper surface of the leaf smooth or nearly so.................................*Celtis iguanaea*
    4. Branches unarmed; flowers and fruit solitary or paired
        6. Leaves mostly lanceolate, at least twice as long as broad, usually with a long-
           tapering apex; upper surface of mature leaves pale yellow green, usually smooth
           to the touch...*Celtis laevigata*

6. Leaves mostly ovate, usually less than twice as long as broad, the apex short-tapered, acute, acuminate, or blunt; upper surface of mature leaves medium or dark green, usually scabrous with the feel of fine sandpaper..........*Celtis tenuifolia*

!*Celtis ehrenbergiana* (Klotzsch) Liebmann. Spiny Hackberry, Desert Hackberry. Shell middens. Lee County. State endangered.

!*Celtis iguanaea* (Jacquin) Sargent. Iguana Hackberry. Shell middens. Manatee, Lee, Collier counties. State endangered.

Celtis laevigata Willdenow. Sugarberry, Hackberry. Floodplains, bottomlands, mixed woodlands, bluff and ravine slopes, old fields. Statewide except the Keys.

Celtis tenuifolia Nuttall. Dwarf Hackberry, Georgia Hackberry. Dry or moist slopes and woodlands, fields, margins of limestone glades, bluffs, ravine slopes, often in association with limestone. Panhandle, east to about Alachua County.

!*Trema lamarckiana* (Schultes) Blume. Pain-in-the-Back, West Indian Trema, Lamarck's Trema. Tropical hammocks, shell middens. Southern peninsula, Collier, Miami-Dade, and Monroe counties, including the Keys. State endangered.

Trema micrantha (Linnaeus) Blume. Nettle Tree. Hammocks, disturbed sites. Southern peninsula, Pinellas and Martin counties southward along the coast, inland south of Lake Okeechobee.

*Trema orientalis (Linnaeus) Blume. African Elm, Mozambique Trema. Disturbed sites, escaped from cultivation. Martin, Palm Beach, Miami-Dade counties. Asia, Africa, and Australasia.

Capparaceae (Caper Family)

1. Lower surface of leaves and upper stem conspicuously covered with small scales (seen best with 10× magnification); stamens usually not exceeding about 14 in number..*Capparis cynophallophora*
1. Lower surface of leaves lacking scales; stamens usually exceeding 24 in number.........
   ....*Capparis flexuosa*

Capparis cynophallophora Linnaeus. Jamaican Caper. Coastal hammocks, shell middens. Southern peninsula from about Pinellas and Brevard counties southward, especially along the coast.

Capparis flexuosa (Linnaeus) Linnaeus. Bay-leaved Caper, Limber Caper. Coastal hammocks, shell middens. East coast and southern peninsula from about Collier and Volusia counties southward, essentially along the coast.

Caprifoliaceae (Honeysuckle Family)

1. Plant a vine
    2. Flowers red; uppermost pair of leaves below the flowers lacking a petiole, the bases fused and encircling the stem (perfoliate)...............................*Lonicera sempervirens*
    2. Flowers white, creamy yellow, or pinkish; uppermost pair of leaves petiolate............
    ....*Lonicera japonica*
1. Plant a shrub
    3. Flowers in conspicuous, upright, terminal, cymose clusters.........*Abelia × grandifolia*
    3. Flowers in inconspicuous, dangling, axillary spikes........*Symphoricarpos orbiculatus*

*****Abelia × grandiflora (André) Rehder.** Largeflower Abelia. Disturbed sites, escaped from cultivation. Franklin and Escambia counties, perhaps elsewhere. Asia.
*****◆*Lonicera japonica* Thunberg.** Japanese Honeysuckle. Disturbed sites, hammocks, rich woodlands, ravine slopes, bottomlands, roadsides, fields, thickets. Panhandle southward to Lake County, scattered southward to Miami-Dade County, potentially throughout the state. Asia. FLEPPC listed (I).
Lonicera sempervirens **Linnaeus.** Coral Honeysuckle, Trumpet Honeysuckle. Hammocks, dry uplands, woodland borders, sandy pinelands, roadsides. Panhandle, south to Sarasota and Highlands counties.
Symphoricarpos orbiculatus **Moench.** Coralberry, Indian-currant. Coastal and inland hammocks, usually where limestone is evident. Jackson and Levy counties.

Caricaceae (Papaya Family)

Carica papaya **Linnaeus.** Papaya. Native but widely cultivated and escaped from plantings. Distributed along the east coast as far north as Duval County, more common in south Florida and the Keys.

Casuarinaceae (Beefwood or Sheoak Family)

1. Scalelike leaves 6–8 per node..*Casuarina equisetifolia*
1. Scalelike leaves 7 or more per node
    2. Scalelike leaves 7–10 per node.................................*Casuarina cunninghamiana*
    2. Scalelike leaves 10–17 per node..*Casuarina glauca*

*◆*Casuarina cunninghamiana* **Miquel.** Cunningham's Beefwood, River Sheoak. Disturbed sites. Central and southern peninsula, from about Brevard and Charlotte counties southward. Eastern Australia. FLEPPC listed (II).

*◆*Casuarina equisetifolia* **Linnaeus.** Australian Pine, Horsetail Casuarina. Disturbed sites, coastal dunes, beaches. Along the coast from Franklin and Volusia counties southward, including the Keys. Asia, Indonesia, Australia. FLEPPC listed (I).

*◆*Casuarina glauca* **Sieber ex Sprengel.** Brazilian Beefwood, Gray Sheoak, Australian Pine. Disturbed sites. Pinellas, Seminole, and Brevard counties southward, also Franklin and St. Johns counties. Southeastern Australia. FLEPPC listed (I).

Cecropiaceae (Cecropia Family)

*◆*Cecropia palmata* **Willdenow.** Trumpet Tree. Rockland and tropical hammocks. Tropical America. FLEPPC listed (II).

Celastraceae (Staff Tree Family)

1. Leaves alternate
   2. Plant a vine...*Celastrus paniculatus*
   2. Plant a shrub or small tree
      3. Venation on upper leaf surface distinct, petals and sepals 4, fruit a drupe............
      ....*Schaefferia frutescens*
      3. Venation on upper leaf surface obscure, petals and sepals 5, fruit a capsule.........
      ....*Maytenus phyllanthoides*
1. Leaves opposite
   4. Plant a vine...*Hippocratea volubilis*
   4. Plant a shrub or small tree
      5. Fruit a capsule, seeds covered by a reddish, reddish orange, or pinkish orange aril
         6. Fruit red at maturity, the surface warty or bumpy, petioles, if present, 1–3 mm
            long...*Euonymus americanus*
         6. Fruit pinkish or pinkish purple at maturity, the surface smooth, petioles at least
            1 cm long
            7. Undersurface of leaf with erect hairs, petiole 8–20 mm long, fruit pinkish or
               purplish, flower petals maroon or green suffused with maroon, new branches
               green, older branches brown...................................*Euonymus atropurpureus*
            7. Undersurface of leaf mostly glabrous (a few hairs on the major veins),
               petiole 5–33 mm long, fruit pinkish, pinkish red, yellowish, or yellowish

brown, flower petals white, branches green throughout.....*Euonymus maackii*
5. Fruit a drupe, seed lacking a colorful covering (aril)
    8. Leaf margins predominately crenate throughout or with spine-tipped teeth
        9. Plant a low-growing shrub, marginal leaf teeth sharp, spinulose-toothed.......
        ....……*Crossopetalum ilicifolium*
        9. Plant larger, leaf margins crenulate...................…....*Crossopetalum rhacoma*
    8. Margins of many leaves entire, at least proximally, those of few to many leaves
      very bluntly toothed distally...*Gyminda latifolia*

*Celastrus paniculatus **Willdenow.** Oriental Bittersweet, Staff-Vine. Escaped from culti-
vation in southern Florida, primarily Miami-Dade County. Asia.
!Crossopetalum ilicifolium **(Poiret) Kuntze.** Christmasberry. Pinelands. Collier and
Miami-Dade counties, including the Keys. State threatened.
!Crossopetalum rhacoma **Crantz.** Maidenberrry, Rhacoma, Florida Crossopetalum.
Hammocks, pinelands. Primarily Miami-Dade County and the Keys, also collected in a
remnant hammock in Sarasota County. State threatened.
Euonymus americanus **Linnaeus.** American Strawberrybush, Hearts-a-bustin'-with-love.
Moist woods and hammocks, floodplains. Throughout the panhandle, south to Highlands
and Sarasota counties.
!Euonymus atropurpureus **Jacquin.** Eastern Wahoo, Burning Bush. Slopes, ravines,
mesic forests, mostly in association with the Apalachicola River. Gadsden and Liberty
counties. State endangered.
*Euonymus maackii **Ruprecht.** Spindletree. Hammocks. Collected in St. Johns County.
China.
!Gyminda latifolia **(Swartz) Urban.** West Indian False Box. Subtropical hammocks.
Southernmost peninsula, essentially Miami-Dade County and the Keys. State endangered.
Hippocratea volubilis **Linnaeus.** Medicine Vine. Subtropical hammocks. Southern pen-
insula, St. Lucie and Lee counties southward, including the Keys.
!Maytenus phyllanthoides **Bentham.** Florida Mayten. Dunes, coastal hammocks. West
coast, Levy County south to Miami-Dade County, including the Keys. State threatened.
!Schaefferia frutescens **Jacquin.** Florida Boxwood. Subtropical hammocks. Miami-Dade
County and the Keys. State endangered.

Chrysobalanaceae (Coco Plum Family)

1. Plant typically low-growing, usually less than 50 cm tall, rarely 1–2 m tall; inflorescence terminal..*Licania michauxii*
1. Plant typically a dense shrub or small tree greater than 50 cm tall; flowers axillary........ ....*Chrysobalanus icaco*

Chrysobalanus icaco **Linnaeus.** Coco Plum. Low coastal hammocks, beaches, sand dunes, cypress heads, widely cultivated along roadsides and in residential yards. Southern peninsula, Pasco and Brevard counties southward, especially along the coast.
Licania michauxii **Prance.** Gopher Apple. Sandhills, sandy roadsides, dry pinelands. Statewide.

Clethraceae (White Alder Family)

Clethra alnifolia **Linnaeus.** Sweet Pepperbush, Summersweet. Swamps, bays, bogs, moist flatwoods, ditch banks. Largely restricted to the panhandle from Escambia County to the Suwannee River, also Nassau, Putnam, and Lake counties.

Clusiaceae (The Garcinia Family)

1. Flowers solitary or in terminal branches; leaves obovate; fruit many-seeded.................. ....*Clusia rosea*
1. Flowers in axillary racemes; leaves elliptic; fruit a single-seeded drupe
    2. Stamens 40–50; fruit 2–2.5 cm long.....................*Calophyllum antillanum*
    2. Stamens 200–300; fruit 2.5–4 cm long...............................*Calophyllum inophyllum*

◆Calophyllum antillanum **Britton.** Santa Maria, Galba. Mangrove swamps, disturbed hammocks, escaped from cultivation. Martin County to Miami-Dade County and the Keys. West Indies. FLEPPC listed (I).
Calophyllum inophyllum **Linnaeus.** Alexandrian Laurel. Disturbed sites. Miami-Dade County. Southeast Asia.
Clusia rosea **Jacquin.** Autograph Tree, Pitch Apple, Balsam Apple. Hammocks, widely cultivated in commercial and residential landscapes. Broward and Miami-Dade counties and the Keys.

Combretaceae (Combretum Family)

1. Leaves opposite
   2. Plant a tree or shrub, flowers greenish, inconspicuous............*Laguncularia racemosa*
   2. Plant a shrub or vine, flowers white, red, or pink, conspicuous........*Quisqualis indica*
1. Leaves alternate
   3. Leaves predominately not exceeding 10 cm long and not exceeding 4 cm wide
      4. Flowers in rounded heads; fruit a dry, button- or conelike head, evident most of the year
         5. Leaves green...*Conocarpus erectus*
         5. Leaves grayish or silvery.................................*Conocarpus erectus* var. *sericeus*
      4. Flowers in short or long racemes or axillary clusters; fruit not as above
         6. Leaves thick, more or less succulent; flowers with 5 white, spreading petals, each to about 5 mm long, borne in axillary clusters..........*Lumnitzera racemosa*
         6. Leaves not succulent; flowers greenish white, petals not evident, borne in short or elongated spikes from the leaf axils
            7. Flowers in elongated spikes; leaves 3–9 cm long; plants lacking (or with very few) spines...*Bucida buceras*
            7. Flowers in short spikes; leaves not exceeding 3 cm long; plants conspicuously spiny...*Bucida molinettii*
   3. Leaves predominately greater than 10 cm long and greater than 4 cm broad
      8. Leaves long-elliptic or lanceolate, apex pointed........................*Terminalia arjuna*
      8. Leaves obovate, broadly elliptic, or broadly spatulate, apex rounded
         9. Fruit winged or angled; many leaves greater than 15 cm long.......................
         ....*Terminalia catappa*
         9. Fruit not winged or angled; virtually all leaves not exceeding 15 cm long............
         ....*Terminalia muelleri*

Bucida buceras **Linnaeus.** Black Olive. Cultivated and escaped in southern Florida. From about Broward County southward on the east coast, including the Keys; also naturalized in Charlotte County. West Indies.

Bucida molinetii **(M. Gómez) Alwan & Stace.** Spiny Black Olive. Hammocks. Very rare in Miami-Dade County.

Conocarpus erectus **Linnaeus.** Buttonwood. Tidal swamps, margins of mangrove swamps, cultivated inland. Mainly coastal from Levy and Volusia counties southward, throughout the southernmost peninsula, including the Keys.

Conocarpus erectus **Linnaeus var.** *sericeus* **Grisebach.** Silver Buttonwood. Similar habitats as buttonwood, often cultivated along roadsides and in residential and commercial landscapes.

Laguncularia racemosa **(Linnaeus) C. F. Gaertner.** White Mangrove. Tidal swamps. Mainly coastal from Levy and Volusia counties southward, through the southernmost peninsula, including the Keys.

◆Lumnitzera racemosa **Willdenow.** Black Mangrove. Disturbed sites, mangrove

swamps. Miami-Dade County. Southeast Asia, eastern Africa, northern Australia. Potentially invasive. FLEPPC listed (II).

*__Quisqualis indica__ **Linnaeus.** Rangoon Creeper. Disturbed sites. Highlands, Broward, and Miami-Dade counties. Southeast Asia.

*__Terminalia arjuna__ **(Roxburgh) Beddome.** Wild Almond. Miami-Dade County. India and Ceylon.

*◆__Terminalia catappa__ **Linnaeus.** West Indian Almond, Sea-Almond, Tropical Almond. Cultivated for ornament, escaped, sandy sites and along beaches throughout southeast Florida and the Keys, northward to at least Brevard County. Asia. FLEPPC listed (II).

*◆__Terminalia muelleri__ **Bentham.** Australian Almond. Mangrove swamps, hammocks, disturbed woodlands. Southern peninsula, from about Manatee and Palm Beach counties, southward. Australia. FLEPPC listed (II).

Convolvulaceae (Morning-Glory Family)

1. Corolla white with a dark center at base inside; leaves heart shaped, entire; fruit 1- or 2-seeded...*Turbina corymbosa*
1. Corolla lacking a dark center at base inside, or if with a dark center, leaves deeply lobed; fruit several seeded
   2. Stem and leaves with branched hairs, stigmas 2
      3. Flowers blue or pinkish
         4. Flowers aggregated into a dense cluster and subtended by conspicuous, hairy leaflike bracts...*Jacquemontia tamnifolia*
         4. Flowers solitary, borne at the tip of an elongated pedicel, lacking leaflike bracts..*Jacquemontia pentanthos*
      3. Flowers white
         5. Margins of outer sepals lined with short hairs, or outer sepals sparsely hairy throughout..*Jacquemontia reclinata*
         5. Margins of outer sepals lacking hairs, or hairs, if any, tufted at the apex
           6. Apex of outer sepals rounded............................*Jacquemontia curtisii*
           6. Apex of outer sepals acute............................*Jacquemontia havanensis*
   2. Stem and leaves with simple hairs or lacking hairs, stigma 1
      7. Flowers white
         8. Flowers with a red or dark center at base inside, leaves deeply 5–9 lobed, lobes toothed..*Merremia dissecta*
         8. Flowers with a yellow center at base inside, leaves heart-shaped, sometimes shallowly lobed at base..*Ipomoea alba*
      7. Flowers yellow or lavender
         9. Flowers yellow...*Merremia tuberosa*
         9. Flowers lavender, deep pink, or purple, usually with a darker throat............... ....*Ipomoea carnea* subsp. *fistulosa*

Ipomoea alba **Linnaeus.** Moonflower, Tropical White Morning-Glory. Mangrove margins, hammocks. Duval and Levy counties southward, perhaps farther north along the west coast.

*◆Ipomoea carnea **Jacquin subsp.** *fistulosa* **(Martius ex Choisy) D. F. Austin.** Bush Morning-Glory. Disturbed sites. Gulf coast, Pinellas County south to the Keys. South America. FLEPPC (II).

•!Jacquemontia curtisii **Peter ex Hallier f.** Pineland Clustervine, Pineland Jacquemontia. Pinelands, pine rocklands. Southern peninsula, Martin and Hendry counties southward, not including the Keys. State threatened.

!Jacquemontia havanensis **(Jacquin) Urban.** Havana Clustervine. Hammocks. Keys. State endangered.

!Jacquemontia pentanthos **(Jacquin) G. Don.** Skyblue Clustervine. Hammocks. Broward and Collier counties southward, including the Keys. State endangered.

•!Jacquemontia reclinata **House ex Small.** Beach Clustervine, Beach Jacquemontia. Pineland, dunes. Southeast peninsula, Martin to Miami-Dade counties. State endangered.

Jacquemontia tamnifolia **(Linnaeus) Grisebach.** Hairy Clustervine. Disturbed sites, fallow agricultural fields, residential gardens. Scattered statewide, not including the Keys.

*Merremia dissecta **(Jacquin) Hallier f.** Noyau Vine, Wood Rose, Alamo Vine. Disturbed sites, roadsides, medians, fallow fields, woodland margins. Throughout the state, not including the Keys. Tropical America, considered native to Florida by some observers.

*◆Merremia tuberosa **(Linnaeus) Rendle.** Spanish Arborvine, Yellow Morning-Glory. Disturbed sites, hammock margins. Southeast peninsula, Broward County south to the Keys; also reported escaped from cultivation in Hillsborough County. Mexico. FLEPPC listed (II).

*Turbina corymbosa **(Linnaeus) Rafinesque.** Christmasvine. Hammocks. Southern peninsula, Palm Beach and Lee counties southward including the Keys. Mexico south to Peru.

Cornaceae (Dogwood Family)

1. Leaves, at least those at mid-branch, conspicuously alternate (those near the end of the branch often closely clustered and not obviously alternate).........*Cornus alternifolia*
1. Leaves opposite
    2. Inflorescence a buttonlike head subtended by 4 large, white, petallike bracts; fruit red..*Cornus florida*
    2. Flowers in an open cyme, not subtended by 4 large, conspicuous white, petallike bracts; fruit not red
        3. Upper surface of leaves smooth to the touch, or nearly so, plants mostly of wetland habitats
            4. Upper surface of leaves glabrous; pith of older twigs brownish...........................
            ....*Cornus amomum*

4. Upper surface of leaves glabrous or with short appressed hairs; pith of older twigs white..*Cornus foemina*

   3. Upper surface of leaves definitely rough to the touch with the feel of fine sandpaper, this from erect, stiff hairs..*Cornus asperifolia*

!*Cornus alternifolia* **Linnaeus f.** Pagoda Dogwood, Alternate-leaved Dogwood. Mesic woodlands, lower slopes of moist ravine slopes. Leon, Gadsden, Calhoun, and Walton counties. State endangered.

Cornus amomum **Miller.** Silky Dogwood. Pond and lake margins, floodplain forests. Central and extreme western panhandle, Jackson to Franklin counties, Escambia County.

Cornus asperifolia **Michaux.** Roughleaf Dogwood. Upland forests, mesic hammocks, often where limestone is present at or near the surface, occasionally in bottomlands. Extreme western panhandle, east to Nassau and Duval counties, south to Hernando and Sumter counties.

Cornus florida **Linnaeus.** Flowering Dogwood. Well-drained upland woods of various mixtures, often planted in parks, residential lawns, and along city streets. Panhandle east to Duval County, south to Manatee and Polk counties.

Cornus foemina **Miller.** Stiff Cornel Dogwood. Swamp margins, margins of large streams, often in standing water. Northern Florida, south on the west coast to Monroe County, on the east coast to Martin County.

Cupressaceae (Cedar Family)

1. Leaves alternate, plant deciduous
   2. Leaves of median and lower branches of mature trees appressed to the twig and overlapping, not 2-ranked, bark with visible furrows, main branches mostly ascending, plant usually growing in ponded water.....................................
     ....*Taxodium ascendens*
   2. Leaves of median and lower branches of mature trees spreading, featherlike, 2-ranked, bark mostly smooth or with very shallow furrows, main branches mostly spreading, plant usually growing in flowing water......................................
     ....*Taxodium distichum*
1. Leaves opposite or whorled, plant evergreen
   3. Ovulate cones superficially berrylike, fleshy, bluish, the surfaces smooth, remaining closed at maturity..*Juniperus virginiana*
   3. Cones not berrylike, hard, woody, surfaces roughened, the scales separating at maturity to release the seeds
     4. Mature leaves 3 per whorl, branchlets rounded in cross-section............
     ....*Callitris glaucophylla*
     4. Mature leaves opposite, leaves in 4 ranks, branchlets flattened in cross-section

5. Ovulate cones with flat scales, more or less egg-shaped or ellipsoid, branches usually vertically disposed.....................................*Platycladus orientalis*
5. Ovulate cones with peltate scales, more or less round, branches usually horizontally disposed
    6. Nearly all leaves with a tiny, resinous gland on the back............................
    ....*Chamaecyparis thyoides* var. *thyoides*
    6. Leaves on principal stems of a branch system with a resinous gland on the back, other leaves lacking glands........*Chamaecyparis thyoides* var. *henryae*

*Callitris glaucophylla **Joy Thompson & L. A. Johnson.** White Cypress-pine. Scrub. Central peninsula, Orange and Brevard south to Lee and Broward counties.

Chamaecyparis thyoides **(Linnaeus) Britton et al. var.** *thyoides.* Swamps, bogs, floodplains, spring runs, usually in association with moving water. Sporadically distributed in the panhandle and northern peninsula, Escambia to Lake counties.

Chamaecyparis thyoides **(Linnaeus) Britton et al. var.** *henryae* **(H. L. Li) Little.** Atlantic White Cedar. Swamps, bogs, floodplains. Western panhandle, about Walton County westward.

Juniperus virginiana **Linnaeus.** Red Cedar. Coastal and inland hammocks, beaches, shell mounds, limestone glades, occurring naturally mostly in calcareous soils, but now expanded into other soil types. Potentially throughout the state, not including the Keys; more common in the panhandle and north and south-central peninsula.

*Platycladus orientalis **(Linnaeus) Franco.** Oriental Arborvitae. Escaped from cultivation, disturbed areas, mostly coastal. Marion and Brevard counties, potentially elsewhere. Asia, mainly China.

Taxodium ascendens **Brongniart.** Pond-cypress. Swamps, depression ponds, lake margins, usually in association with non-moving water. Nearly throughout the state; absent from the Keys.

Taxodium distichum **(Linnaeus) Richard.** Bald-cypress. Floodplains, drainages, lake margins, swamps, usually in association with moving water. Nearly throughout the state; absent from the Keys.

Cycadaceae (Cycad Family)

*Cycas revoluta **Thunberg.** Sago Palm. Weedy disturbed sites, escaped from cultivation, commonly planted in residential and commercial landscapes. Reported naturalized in Escambia and Pasco counties; potentially naturalized elsewhere.

Cyrillaceae (Titi Family)

1. Lateral leaf veins obscure, stamens 10, fruit 2–5-winged (remaining on most plants nearly all year), bark dark gray or blackish, flowering in early spring..........................
....*Cliftonia monophylla*
1. Lateral leaf veins evident, stamens 5, fruit not winged, bark reddish or gray, flowering in early summer...*Cyrilla racemiflora*

***Cliftonia monophylla* (Lamarck) Britton ex Sargent.** Black Titi, Buckwheat Tree. Swamps, bayheads, shrub bogs, wet flatwoods, flatwoods depressions, stream margins in acid woods. Panhandle, east to about Jefferson County, disjunct to Clay, Putnam, and Marion counties.
***Cyrilla racemiflora* Linnaeus.** Swamp Cyrilla, Titi. Bayheads, wet swamps, shrub bogs, stream courses, margins of alluvial rivers, often in standing water. Throughout northern Florida, south to about Highlands County. A form with leaves 1–4 cm long and occurring in scrub and on inland dunes in south-central Florida has been called *C. arida* Small. A form with very small leaves and short racemes (< 4 cm long) distributed chiefly along the panhandle coast is sometimes referred to *C. parvifolia* Rafinesque-Schmaltz.

Ebenaceae (Ebony Family)

1. Leaf moderately thin in textrue, lower surface glabrous or with at least a few hairs, the apex acuminate...*Diospyros virginiana*
1. Leaf thick, both surfaces definitely glabrous, the apex usually rounded or bluntly pointed
    2. Leaf blade with a pair of glands at the base................................*Diospyros maritima*
    2. Leaf blade lacking glands at the base...*Diospyros ebenum*

***Diospyros ebenum* J. König ex Retzius.** Ebony. Disturbed woodlands, escaped from cultivation. Broward and Palm Beach Counties. Southern India, Sri Lanka.
***Diospyros maritima* Blume.** Disturbed sites, escaped from cultivation. Miami-Dade County. Tropical western Pacific, western Pacific islands, Australia.
***Diospyros virginiana* Linnaeus.** Common Persimmon. Upland woodlands of various mixtures from dry sandhill pinelands to moist bottomlands. Throughout except the Keys.

Elaeagnaceae (Oleaster Family)

1. Lower surface of leaf with numerous silvery scales that are in turn dotted with fewer brownish scales; branches of the season usually producing axillary thorns; leaves evergreen...*Elaeagnus pungens*
1. Lower surface of leaf with numerous silvery scales but few or no brownish scales; branches of the season usually not producing axillary thorns; leaves deciduous.............. ....*Elaeagnus umbellata*

*◆*Elaeagnus pungens* **Thunberg.** Silver Thorn, Thorny Olive. Disturbed sites, rich mesic forests, often in dry sandy areas, escaped from cultivation. Scattered across the panhandle and northern peninsula, south to about Marion County, also reported in St. Lucie County. Japan. FLEPPC listed (II).
*◆*Elaeagnus umbellata* **Thunberg.** Silverberry, Autumn Olive. Disturbed sites, escaped from cultivation. Leon and Gadsden counties. Asia. FLEPPC listed (II).

Ericaceae (Heath Family)

1. Leaves needlelike, flowers tiny...*Ceratiola ericoides*
1. Leaves not needlelike, flowers larger
  2. Fruit fleshy
    3. Fruit a 10-seeded drupe, lower surface of leaf usually with conspicuous, distinctive amber-colored glands (these seen best with magnification)
      4. Bracts of the inflorescence equal in length to or longer than the pedicels; pedicels with gland-tipped hairs
        5. Corolla often less than 5 mm long, plants usually less than 40 cm tall, young twigs, axis of inflorescence, pedicels, and floral tube with short, curly hairs, some of which may be gland-tipped, lower surface of leaf with amber-colored glands..*Gaylussacia dumosa*
        5. Corolla often greater than 5 mm long, plants often greater than 40 cm tall, young twigs, axes of the inflorescences, pedicels, and floral tube with long, spreading, silvery, minutely gland-tipped hairs, glands of lower surface of leaf obscure, reddish...*Gaylussacia mosieri*
      4. Bracts of the inflorescence shorter than the pedicels; pedicels lacking gland-tipped hairs
        6. Twigs of the season glabrous or nearly so.....................*Gaylussacia frondosa*
        6. Twigs of the season conspicuously hairy
          7. Lower surface of leaf sparsely hairy, strongly glaucous and silvery gray;

larger leaves 2–4 cm long; plant more or less columnar with ascending branches..*Gaylussacia nana*
>> 7. Lower surface of leaf sometimes densely hairy, not glaucous; larger leaves 3–6 cm long; plant loosely branched, branches spreading..........................
....*Gaylussacia tomentosa*
3. Fruit a many-seeded berry, lower surface of leaf lacking amber-colored glands
>> 8. Corolla open at anthesis, the width of the opening about as wide as the diameter of the corolla
>>> 9. Corolla shorter than the stamens at maturity, the anthers conspicuously protruding beyond the tips of the petals.......................*Vaccinium stamineum*
>>> 9. Anthers not protruding well beyond the tip of the petals.............................
....*Vaccinium arboreum*
>> 8. Corolla urceolate at anthesis, the opening constricted, usually narrower than the diameter of the corolla
>>> 10. Plant evergreen, leaves predominantly 5–15 mm long, rarely longer
>>>> 11. Foliage more or less glaucous or bluish green throughout, floral tube and fruit conspicuously so; lower surface of leaves glabrous.........................
....*Vaccinium darrowii*
>>>> 11. Foliage lustrous green throughout, floral tube and fruit slightly if at all glaucous; lower surface of some (often few) leaves with gland-tipped hairs...*Vaccinium myrsinites*
>>> 10. Plant deciduous or evergreen, leaves predominantly greater than 20 mm long
>>>> 12. Larger leaves usually less than 35 mm long, margins finely toothed, the teeth gland-tipped.......................................*Vaccinium elliottii*
>>>> 12. Larger leaves greater than 35 mm long, margins entire, or if toothed, the teeth not gland-tipped
>>>>> 14. Lower surface of leaf and leaf margins with gland-tipped hairs.........
....*Vaccinium virgatum*
>>>>> 14. Lower surface of leaf and leaf margins lacking gland-tipped hairs
>>>>>> 15. Twigs and lower leaf surface with dingy, brownish, or more or less dark pubescence, berry black................*Vaccinium fuscatum*
>>>>>> 15. Twigs and lower surface of leaves glabrous, or if lower surface of leaves hairy, the hairs whitish or transparent and confined to the distal portion of the blade and central vein, berry dark blue and glaucous...*Vaccinium formosum*
2. Fruit not fleshy, often a hard capsule
> 16. Corolla and calyx 7-parted, petals distinctly separate................*Bejaria racemosa*
> 16. Corolla and calyx 5-parted, petals at least partially united, at least at base
>> 17. Plant a trailing or climbing vine or sub-shrub, or low-growing and less than 20 cm tall
>>> 18. Leaves lanceolate with conspicuous whitish veins.....*Chimaphila maculata*
>>> 18. Leaves otherwise
>>>> 19. Plant vinelike, trailing or creeping, stem rooting in the substrate, leafy

branches only slightly elevated above the ground...........*Epigaea repens*
    19. Plant erect or upright, if climbing, usually doing so under the bark of
      Chamaecyparis thyoides or *Taxodium distichum*.....*Pieris phyllyreifolia*
17. Plant a definite shrub or tree, never vinelike, much taller than 20 cm at
  maturity
    20. Blades of mature leaves at least 15 cm long and 2.5 cm broad; plant a
      single-stemmed tree...*Oxydendrum arboreum*
    20. Blades of mature leaves much shorter than 15 cm long; plant most often
      with the habit of a small or large shrub, rarely single-stemmed and
      arborescent
      21. Corolla campanulate bell-shaped, or tubular and spreading at the
        apex, the aperture broad and open in comparison to the width of the
        body of the corolla
        22. Corolla campulate or bell-shaped
          23. Plant small, usually not exceeding 60 cm tall, perhaps to 1 m
            tall, leaves 5–15 mm long, 2–8 mm broad..........*Kalmia hirsuta*
          23. Plant potentially large, usually greater than 1 m tall,
            potentially to 9 m tall, leaves 2–10 cm long, 3–5 cm broad.......
            ....*Kalmia latifolia*
        22. Corolla tubular, flaring at the apex, the aperture open
          24. Leaves evergreen, lower surface with small, rust-colored
            scales; stamens 10............*Rhododendron minus* var. *chapmanii*
          24. Leaves deciduous, lower surface lacking rust-colored scales;
            stamens 5
            25. Flowers orange or yellow.............*Rhododendron austrinum*
            25. Flowers white or pink
              26. Flowers very slender, appearing in summer well after
                leaf expansion.........................*Rhododendron viscosum*
              26. Flowers wider, appearing in spring, before or with leaf
                expansion
                27. Corolla pink, or white blushed with pink, pedicel
                  and calyx lacking glands...
                  ....*Rhododendron canescens*
                27. Corolla white, usually with a yellow blotch at the
                  base of the upper petal; pedicel, calyx, and corolla
                  tube with stalked glands...
                  ....*Rhododendron alabamense*
      21. Corolla urceolate or urn-shaped, the aperture constricted and usually
        narrower than the body of the corolla, or corolla more or less cylindric
        28. Flower stalks with 2 opposite or nearly opposite, narrow, awl-
          shaped bractlets a little below the calyx, not lower than about
          midway the flower stalk.................................*Pieris phyllyreifolia*
        28. Flower stalks with a pair of ovate bractlets, if immediately below
          the calyx, then alternate, or if narrow, then borne at the base of the

flower stalk

29. Pith diaphragmed and chambered..............*Agarista populifolia*

29. Pith solid

    30. Capsule with thickened, whitish ridges over the sutures, these usually falling after the capsule has split

        31. Lower surface of the leaf blade, flower stalk, and calyx with small scales (requires at least 10× magnification to see clearly)

            32. Ultimate branches spreading, flowers nearly always restricted to wood of the previous season, leaves at the branch tips usually more or less uniform in size with those below, upper surface of leaves usually green or rust-colored, margins often revolute...........

                ....*Lyonia ferruginea*

            32. Ultimate branches usually rigidly ascending, flowers nearly always borne on new wood of the season, leaves at the branch tips usually smaller than those below, upper surface of leaves often with a bluish or grayish cast, especially early in the season, margins usually flat or only slightly revolute..

                ....*Lyonia fruticosa*

        31. Lower surface of leaf blade, flower stalk, and calyx glabrous or hairy, but lacking scales

            33. Leaf margins finely toothed, corolla urn-shaped, less than 5 mm long.....................*Lyonia ligustrina*

            33. Leaf margins entire, corolla cylindric, greater than 5 mm long

                34. Leaves persisting through winter, blade outlined with a conspicuous, thickened, submarginal vein, corolla averaging less than 5 mm wide.....................................*Lyonia lucida*

                34. Leaves deciduous in winter, blade lacking a thickened submarginal vein, corolla averaging greater than 5 mm wide.............*Lyonia mariana*

    30. Capsule lacking thickened, whitish ridges over the sutures

        35. Leaves persistent though winter, thick; bractlets positioned at the base of the flower stalk................*Leucothoe axillaris*

        35. Leaves deciduous in winter, thin and papery; bractlets of the flower stalk alternate, positioned just below the calyx...

        ....*Eubotrys racemosa*

***Agarista populifolia* (Lamarck) Judd.** Florida Leucothoe, Florida Hobblebush, Pipe-stem, Agarista. Wet woods, moist hammocks, moist uplands, sometimes in association with limestone. Northeast peninsula, Duval County south to Lake and Orange counties.

Bejaria racemosa **Ventenat.** Tarflower. Mesic flatwoods, coastal pinelands. Peninsula, Taylor and Nassau counties south to Collier and northern Miami-Dade counties, absent from the Keys.

Ceratiola ericoides **Michaux.** Florida Rosemary, Sand Heath. Dunes and scrub. Coastally nearly throughout, Escambia to Collier on the west coast, Duval to Miami-Dade on the east coast, central peninsula scrub.

Chimaphila maculata **(Linnaeus) Pursh.** Spotted Wintergreen, Striped Prince's Pine. Hammocks, dry woodlands. Uncommon, Leon County.

!*Epigaea repens* **Linnaeus.** Trailing Arbutus. Rich woods, moist and dry slopes, ravines. Western and central panhandle, Escambia to Okaloosa counties, Liberty County. State endangered.

Eubotrys racemosa **(Linnaeus) Nuttall.** Swamp Doghobble. Swamps, margins of pineland drainages, flatwoods, river banks. Panhandle and northern peninsula, Escambia to Nassau and south to Polk and Osceola counties.

Gaylussacia dumosa **(Andrews) Torrey & A. Gray.** Dwarf Huckleberry. Well-drained pinelands, dry to mesic flatwoods, pine—oak woods, sandhills, scrub, rarely along swamp margins. Panhandle and peninsula, south to Collier and Palm Beach counties.

Gaylussacia frondosa **(Linnaeus) Torrey & A. Gray. Dangleberry.** Pinelands. Central panhandle.

Gaylussacia mosieri **Small.** Wooly Huckleberry. Bogs, wet flatwoods, swamp margins, wet savannas. Panhandle, Escambia to Madison and Lafayette counties, disjunct to Duval and Volusia counties.

Gaylussacia nana **(A. Gray) Small.** Dwarf Dangleberry. Moist pinelands and flatwoods, sandhill pinelands; margins of savannas, bogs, bays, and depressional wetlands. Panhandle, south to central peninsula.

Gaylussacia tomentosa **(A. Gray) Small.** Hairy Dangleberry. Wet and well-drained pinelands, margins of bays and bogs. Panhandle to central peninsula.

Kalmia hirsuta **Walter.** Hairy Wicky, Wicky, Hairy Laurel. Flatwoods, moist and dry pinelands, usually in association with acid soil. Panhandle, south to Hernando and Lake counties.

!*Kalmia latifolia* **Linnaeus.** Mountain Laurel. Rich forested bluffs and lower ravine slopes, creek bottoms. Panhandle, Leon County westward, disjunct to Suwannee County. State threatened.

Leucothoe axillaris **(Lamarck) D. Don.** Swamp Doghobble. Creek bottoms, floodplains of small streams, moist hammocks. Panhandle from Leon and Wakulla counties westward, intermittently eastward to Nassau and Duval counties, south along the central peninsula to Polk County.

Lyonia ferruginea **(Walter) Nuttall.** Rusty Staggerbush. Moist to scrubby flatwoods, dry acid uplands, scrub. Okaloosa County, east to Nassau County, south to Manatee and Glades counties.

Lyonia fruticosa **(Michaux) G. S. Torrey.** Coastal Plain Staggerbush. Moist and dry flatwoods, scrub. Gadsden, Liberty, and Franklin counties, throughout the peninsula, south to Miami-Dade County, absent from the Keys.

Lyonia ligustrina **(Linnaeus) de Candolle var.** *foliosiflora* **(Michaux) Fernald.** Male-

berry. Flatwoods, bog margins. Okaloosa to Nassau counties, south to Glades County.
***Lyonia lucida* (Lamarck) K. Koch.** Fetterbush. Flatwoods and margins of flatwoods depressions, creek bottoms, margins of bogs and cypress wetlands. Throughout except the Keys.
***Lyonia mariana* (Linnaeus) D. Don.** Piedmont Staggerbush. Wet flatwoods, creek bottoms, shrubby wetlands. Gadsden and Liberty counties, east to Duval County, south to Orange, Lake, Hillsborough, and Manatee counties.
***Oxydendrum arboreum* (Linnaeus) de Candolle.** Sourwood. Bluff forests and ravine slopes, sandy coastal hammocks, margins of bayheads. Panhandle, Madison County westward.
***Pieris phyllyreifolia* (Hooker) de Candolle.** Climbing Wicky, Fetterbush. Moist or wet flatwoods, titi-dominated depressions, cypress and white cedar wetlands, bay swamps. Santa Rosa east to Columbia County, south to Lake County.
!*Rhododendron alabamense* Rehder. Alabama Azalea. Upland woods, moist hammocks. Leon and Jefferson counties. State endangered.
!*Rhododendron austrinum* (Small) Rehder. Florida Flame Azalea, Florida Azalea, Orange Azalea. Ravine slopes and bottoms, bluff forests, margins of floodplains, rich moist woods along small streams. Gadsden, Liberty, and Franklin counties westward. State endangered.
***Rhododendron canescens* (Michaux) Sweet.** Piedmont Azalea, Pinxter Azalea, Sweet Pinxter Azalea, Mountain Azalea. Swamps and swamp margins, bog margins, wet flatwoods, lower ravine slopes, creek bottoms, hammocks. Panhandle, east to Nassau County, south to Marion County.
!*Rhododendron minus* Michaux var. *chapmanii* (A. Gray) W. H. Duncan & Pullen. Chapman's Rhododendron. Coastal and inland flatwoods. Gadsden and Leon Counties, south to Gulf and Franklin counties, disjunct to Clay County where probably persisting from cultivation. State and federally endangered.
***Rhododendron viscosum* (Linnaeus) Torrey.** Swamp Azalea. Wet or moist acid flatwoods, swamps, swamp margins. Panhandle, east to Nassau County, south to DeSoto and Highlands counties.
***Vaccinium arboreum* Marshall.** Sparkleberry, Farkleberry. Dry hammocks, flatwoods, sandhill pinelands, scrub. Panhandle and peninsula south to Lee, Hendry, and Martin counties.
***Vaccinium darrowii* Camp.** Darrow's Blueberry. Flatwoods, sandhills, sandy pinelands. Panhandle, east to Clay County, south to St. Lucie and Collier counties.
***Vaccinium elliottii* A. W. Chapman.** Elliott's Blueberry. Moist hammocks, pinelands, ravine slopes, rich temperate hardwood forests. Panhandle and northern peninsula, south to central peninsula.
***Vaccinium formosum* H. C. Andrews.** Southern Highbush Blueberry, Swamp Highbush Blueberry. Bogs, blackwater swamps, seepages, depression ponds, well-drained pinelands, swamp margins, live oak hammocks. Panhandle, east to at least Lafayette County.
***Vaccinium fuscatum* Aiton.** Hairy Highbush Blueberry, Black Highbush Blueberry. Margins of cypress ponds, wetland depressions, shrub–tree bogs, ravine slopes, uplands, ravine bottoms, flatwoods. Panhandle and northern peninsula, south at least to Volusia, Polk, and Hernando counties.

Vaccinium myrsinites **Lamarck.** Shiny Blueberry. Sandhills, flatwoods. Essentially throughout, except the Keys.

Vaccinium stamineum **Linnaeus.** Deerberry. Flatwoods, sandhill pinelands, hammocks, margins of floodplain woods. Panhandle and peninsula, south to Lee and Martin counties.

Vaccinium virgatum **Aiton.** Swamp Blueberry, Rabbiteye Blueberry. Swamps, sandhill pinelands, dry uplands, creek bottoms, flatwoods, mesic hardwood forests. Panhandle, east to Nassau County, south to at least Alachua County.

Euphorbiaceae (Spurge Family)

1. Flowers naked (petals and sepals absent), subtended by or contained within a cuplike involucre, female flower usually 1, consisting of a single pistil, male flowers usually several, each with a single stamen and surrounding the female flower (inflorescence a cyathium)
    2. Inflorescence (cyathium) somewhat boat- or shoe-shaped, apex 2-lipped, appearing as a conspicuous, brightly colored spur......................*Pedilanthus tithymaloides*
    2. Inflorescence not as above, not with a spurlike appendage
        3. Stems with conspicuous spines, slightly or conspicuously angled, leaves usually persistent
            4. Stem conspicuously 4-angled......................................*Euphorbia lactea*
            4. Stem not conspicuously 4-angled..................................*Euphorbia milii*
        3. Stems lacking spines, round in cross-section, leaves quickly deciduous.............
        ....*Euphorbia tirucalli*
1. Inflorescence not as above, solitary, spicate, or cymose
    5. Inflorescence a 3-flowered cyme, the terminal or central flower developing and opening first, followed by the two lateral flowers; petals present
        6. Petals and sepals distinct, in two series, the sepals not petallike
            7. Sepals fused, the fruit not splitting at maturity, the petiole with 2 conspicuous, dark reddish glands on the upper side near the point of attachment to the blade
                8. Lower surface of the leaves and petiole with straight unbranched hairs, lobed leaves with glands in the marginal sinuses, petals about 2.5 cm long, fruit 3–5-seeded...*Aleurites fordii*
                8. Lower surface of the leaves and petiole with branched, starlike hairs, lobed leaves lacking glands in the marginal sinuses, petals less than 1 cm long, fruit 1–2-seeded...*Aleurites moluccanus*
            7. Sepals not fused, the fruit splitting at maturity, the petiole lacking conspicuous reddish glands on the upper side near the point of attachment to the blade, or if glands present, then not conspicuous
                9. Flowers green or yellowish white, not showy..................*Jatropha curcas*

9. Flowers red or pink, showy
   10. Leaves deeply 3–11-lobed, stipules fringed along the margin
      11. Leaves 3–5 lobed, surfaces with stalked glands............................
      ....*Jatropha gossypifolia*
      11. Leaves 5–11-lobed, surfaces glabrous..............*Jatropha multifida*
   10. Leaves shallowly lobed, often only at the base, sometimes fiddle-shaped
      ....*Jatropha integerrima*
6. Petals and sepals similar, in one series, the sepals petallike
   12. Leaf segments wavy or lobed, especially toward the apex, calyx of the male
   flowers 12–15 mm long, fruit not winged.....................*Manihot grahamii*
   12. Leaf segments entire, calyx of male flowers less than 10 mm long, fruit
   winged...*Manihot esculenta*
5. Inflorescence a raceme or spike, petals absent or if present, then the sepals not petallike
   13. Petals present in male flowers, even if small and rudimentary
      14. Leaf blade with glands at or near its junction with the petiole, lower surface
      of leaf with branched hairs or peltate scales, sap exuding from broken
      petioles milky
         15. Lower surface of leaves densely covered with peltate scales, the
         appearance silvery with brown or reddish orange flecks
            16. Leaves oblong, linear, or elliptic, male flowers with petals, plants
            usually of sandhill and scrub......................*Croton argyranthemus*
            16. Leaves usually ovate, male flowers lacking petals, plants usually of
            coastal dunes..*Croton punctatus*
         15. Lower surface of leaves with branched hairs, not appearing silvery with
         brown or reddish orange flecks..............................*Croton linearis*
      14. Leaves lacking glands at the junction of the petiole and blade, lower leaf
      surface with straight hairs, sap exuding from broken petioles clear............
      ....*Argythamnia blodgettii*
   13. Petals absent in male flowers
      17. Leaves deeply palmately lobed, the petiole attached near the center of the
      blade (peltate)..*Ricinus communis*
      17. Leaves neither deeply lobed nor peltate
         18. Fruit a several-parted fleshy or "woody" conelike capsule, 8–9 cm
         broad, breaking at maturity into curved segments, trunk and branches
         spiny..*Hura crepitans*
         18. Fruit otherwise, trunk and branches not spiny
            19. Leaves varicolored, often with red or purple spots, style deeply cut
            into narrow, pointed segments.....................*Acalypha amentacea*
            19. Leaves not varicolored, lacking red or purple spots, style undivided
               20. Fruit 6-parted, plants dangerously poisonous, sap producing
               burnlike lesions on the skin.................*Hippomane mancinella*
               20. Fruit 2–3-parted
                  21. Petiole with 2 glands at the apex

22. Leaves about as wide as long, margins entire...............
....*Sapium sebiferum*
22. Leaves elliptic or linear, conspicuously longer than wide,
margins toothed.......................…....*Sapium glandulosum*
21. Petiole lacking glands at the apex
23. Leaf margins serrate with gland-tipped teeth..............…...
....…...........*Stillingia aquatica*
23. Leaf margins entire or dentate, not glandular serrate
24. Floral bracts lacking glands at the base, plants of
tropical hammocks of the southernmost peninsula and
the Keys.............................…...*Gymnanthes lucida*
24. Floral bracts with 2 glands at the base, plants
distributed from the central peninsula northward.......
.....................................…...........*Sebastiania fruticosa*

*Acalypha amentacea **Roxburgh subsp.** *wilkesiana* **(Müller Argoviensis) Fosberg.**
Wilkes' Copperleaf. Moist, disturbed sites, escaped from cultivation. Central peninsula
and the Keys. Tropical America.

*◆Aleurites fordii **Hemsley.** Tungoil Tree. Disturbed sites, roadsides, invasive in upland
woodlands. Panhandle, west and south to Alachua, Marion, and Citrus counties. Central
and western Asia. FLEPPC listed (II).

*Aleurites moluccanus **(Linnaeus) Willdenow.** Candlenut Tree, Indian Walnut. Disturbed sites, escaped from cultivation. Southeastern peninsula, Broward and Miami-Dade
counties. Hawaii, Polynesia, Southeast Asia.

•!Argythamnia blodgettii **(Torrey ex Chapman) Chapman.** Blodgett's Silverbush,
Blodgett's Wild Mercury. Pinelands, tropical hammocks. Miami-Dade County and the
Keys. State endangered.

Croton argyranthemus **Michaux.** Silver Croton, Healing Croton. Sandhills, xeric pinelands, scrub. Panhandle, east to Nassau County, south to Pinellas and Osceola counties.

Croton linearis **Jacquin.** Pineland Cotton, Grannybush. Sandhills. Southeastern peninsula, St. Lucie County south to the Keys.

Croton punctatus **Jacquin.** Gulf Croton, Beach Tea. Dunes. Coastal counties with deep
sand dunes, nearly throughout.

*Euphorbia lactea **Haworth.** Mottled Spurge. Persistent from cultivation, probably not
technically naturalized. Keys. India and the Moluccas.

*Euphorbia milii **Des Moulins.** Crown-of-Thorns, Christplant. Disturbed sites, escaped
from cultivation. Sparsely naturalized in the southern peninsula, potentially from about
Sarasota and Broward counties southward. Madagascar.

*Euphorbia tirucalli **Linnaeus.** Pencil Tree, Indian Tree Spurge. Disturbed sites, escaped
from cultivation. Southwestern peninsula, Manatee County south to the Keys. Africa.

Gymnanthes lucida **Swartz.** Crabwood, Oysterwood. Subtropical hammocks. Southeastern peninsula, Palm Beach County south to the Keys.

!Hippomane mancinella **Linnaeus.** Manchineel. Subtropical hammocks. Southernmost
peninsula, Monroe and Miami-Dade counties, including the Keys. State endangered.

*Hura crepitans **Linnaeus.** Sandboxtree. Subtropical hammocks. Miami-Dade County

and the Keys. Tropical America.

*Jatropha curcas **Linnaeus.** Nutmeg Plant, Physic Nut, Barbados Nut. Disturbed sites, escaped from cultivation. East coast, potentially Brevard to Broward counties. Tropical America, Bemuda.

*Jatropha gossypifolia **Linnaeus.** Bellyache Bush. Disturbed sites, escaped from cultivation. Southwestern peninsula, Pinellas and Hillsborough counties, south to Miami-Dade County. Tropical America.

*Jatropha integerrima **Jacquin.** Peregrina. Disturbed sites, escaped from cultivation. East coast, Brevard to Miami-Dade counties, including the Keys. Cuba.

*Jatropha multifida **Linnaeus.** Coralbush, Coralplant. Disturbed sites, escaped from cultivation. West Indies, Puerto Rico, Tropical America. Polk and Indian River counties, south potentially to Broward County.

*Manihot esculenta **Crantz.** Cassava, Yuca, Manioc. Disturbed sites, escaped from cultivation. Southeastern peninsula, Martin County to the Keys. South America.

*Manihot grahamii **Hooker.** Manihot, Cassava, Graham's Cassava. Disturbed sites, roadsides, escaped from cultivation. Sparsely naturalized from Escambia County east to Alachua and Marion counties, south to Polk County. South America.

Pedilanthus tithymaloides **(Linnaeus) Poiteau subsp.** *smallii* **(Millspaugh) Dressler.** Jacob's Ladder, Devil's Backbone, Redbird Flower. Subtropical hammocks, pine rocklands, disturbed sites. Southern peninsula, Palm Beach and Collier counties southward, excluding the Keys.

*◆*Ricinus communis* **Linnaeus.** Castor Bean, Castor Oil Plant. Roadsides, disturbed sites, pinelands. Naturalized mostly from Alachua and St. Johns counties south to the Keys, also sparsely naturalized in the panhandle. Africa. FLEPPC listed (II).

*Sapium glandulosum **(Linnaeus) Morong.** Milktree. Disturbed sites. Escambia County, perhaps not extant in Florida. West Indies, South America.

*◆*Sapium sebiferum* **(Linnaeus) Roxburgh.** Popcorn Tree, Chinese Tallowtree. Disturbed sites, especially in coastal counties, invading moist hammocks and wetland margins. Potentially throughout, excluding the Keys. China, Japan. FLEPPC listed (I).

Sebastiania fruticosa **(W. Bartram) Fernald.** Gulf Sebastian-bush. Dry and moist slopes, mesic hammocks, stream banks. Panhandle east to Nassau County, south to Hernando and Flagler counties.

Stillingia aquatica **Chapman.** Corkwood. Swamps and swamp margins, wet roadside ditches, marsh edges. Nearly throughout except the northeast peninsula and the Keys.

Fabaceae or Leguminosae (Legume or Bean Family)

1. Plant a climbing, trailing, or scandent vine (or a vinelike shrub)
  2. Leaves evenly pinnate or bipinnate, lacking a terminal segment or leaflet (pinna)
    3. Plants unarmed, leaves pinnate, flowers differentiated into standard, wing, and keel petals, seeds red and black.....................................*Abrus precatorius*
    3. Plants usually armed with conspicuous hooked prickles, leaves bipinnate, flowers lacking differentiated standard, wing, and keel petals, seeds yellow or gray
      4. Fruit lacking prickles.................................*Caesalpinia crista*
      4. Fruit prickly
        5. Stipules present, seeds gray..........................*Caesalpinia bonduc*
        5. Stipules absent, seeds yellow........................*Caesalpinia major*
  2. Leaves unifoliolate, bifoliolate, trifoliolate, or odd pinnate (with a terminal leaflet)
    6. Most leaves unifoliolate, plant a trailing, viney shrub
      7. Fruit circular, ovate, or kidney shaped in outline, leaves thick.......................................*Dalbergia ecastaphyllum*
      7. Fruit oval or oblong in outline, leaves thin, papery.............*Dalbergia brownei*
    6. Leaves trifoliolate, bifoliolate, or odd pinnate
      8. Leaves bifoliolate, the apex of the leaflets deeply and conspicuously notched.......................................*Bauhinia yunnanensis*
      8. Leaves trifoliate or odd pinnate, apex of leaflets not as above
        9. Leaves trifoliolate
          10. Leaflet stalks subtended by a pair of small, stipulelike appendages
            11. Stipulelike appendages conspicuous, persistent; corolla less than 1.5 cm long, reddish purple....................*Pueraria montana* var. *lobata*
            11. Stipulelike appendages inconspicuous, quickly falling and often absent; corolla at least 3 cm long
              12. Flowers yellow, 6–6.5 cm long; fruit 4–6 cm broad.......................................*Mucuna sloanei*
              12. Flowers purple or white, 3–4 cm long; fruit 1–1.5 cm broad.......................................*Mucuna pruriens*
          10. Leaflet stalks not subtended by a pair of small, stipulelike appendages.......................................*Canavalia rosea*
        9. Leaves odd pinnate
          13. Leaflets numbering 7–19
            14. Fruit and ovary glabrous, fruit stalk 5–15 mm long.......................................*Wisteria frutescens*
            14. Fruit and ovary hairy, fruit stalk 15–20 mm long
              15. Leaflets 7–13, corolla 2–2.7 cm long...............*Wisteria sinensis*
              15. Leaflets 13–19, corolla 1.5–2 cm long...........*Wisteria floribunda*
          13. Leaflets numbering 5–7.......................*Apios americana*
1. Plant a tree, shrub, or scandent shrub
  16. Most leaves unifoliolate or reduced to phyllodes

17. Leaves unifoliolate
  18. Plant a sprawling shrub, often with vinelike branches
    19. Fruit circular, ovate, or kidney shaped in outline, leaves thick............
      ....*Dalbergia ecastaphyllum*
    19. Fruit oval or oblong in outline, leaves thin, papery....*Dalbergia brownei*
  18. Plant a tree or upright shrub
    20. Leaf blade nearly as broad as long, cordate or notched at base, apex
      acuminate or deeply and conspicuously notched
      21. Leaf blade heart-shaped, acuminate at the apex......*Cercis canadensis*
      21. Leaf blade cordate at base, apex deeply and conspicuously notched,
        the blade thus appearing 2-lobed
        22. Plant with spines at the base of the leafstalk....*Bauhinia aculeata*
        22. Plant lacking spines at the base of the leafstalk
          23. Stamens 3.....................................*Bauhinia purpurea*
          23. Stamens 5.....................................*Bauhinia variegata*
    20. Leaf blade about half as broad as long, elliptic, tapered to base and apex
      24. Flowers blue, plants of the panhandle.....................................
        ....*Lupinus westianus* var. *westianus*
      24. Flowers pink or rose, plants of the central peninsula....................
        ....*Lupinus westianus* var. *aridorum*
17. Leaves reduced to phyllodes
  25. Leaves predominantly straight, more or less linear-lanceolate.................
    ....*Acacia retinodes*
  25. Leaves predominantly curved, more or less oblong....*Acacia auriculiformis*
16. Most leaves with 2 or more leaflets, not phyllodial
  26. Leaves trifoliolate (with 3 leaflets)
    27. Terminal leaflets of most leaves oval or elliptic, not exceeding 5 cm long,
      usually shorter than 4 cm, fruit not splitting at maturity, usually containing
      a single seed
      28. Calyx lobe longer than the calyx tube................*Lespedeza thunbergii*
      28. Calyx lobe shorter than or about equaling the tube....*Lespedeza bicolor*
    27. Terminal leaflets of larger leaves usually well over 5 cm long
      29. Leaflets elliptic, tapering to base and apex.................*Cajanus cajan*
      29. Terminal leaflets broad at base, abruptly tapering to the apex (more
        or less hastate)...*Erythrina herbacea*
  26. Leaves with more than three leaflets
    30. Leaves pinnate
      31. Leaves evenly pinnate, lacking a terminal leaflet or pinna
        32. Plant a shrub or very small tree
          33. Corolla differentiated into standard, wing, and keel petals, the
            standard inserted behind the lateral petals (papilionaceous)
            34. Corolla greater than 5 cm long............*Sesbania grandiflora*
            34. Corolla less than 3 cm long
              35. Fruit conspicuously 4-winged

36. Corolla red.............................*Sesbania punicea*

36. Corolla yellow.....................*Sesbania drummondii*

35. Fruit not winged

37. Fruit 4–6 cm long, 3–5 mm broad....*Sesbania virgata*

37. Fruit 10–18 cm long, 3–5 mm broad........................
....*Sesbania sericea*

33. Corolla not differentiated into standard, wing, and keel petals, the lateral petals inserted behind the standard

38. Stalks of individual flowers (pedicels) with tiny but conspicuous bracteoles....*Chamaecrista lineata* var. *keyensis*

38. Stalks of individual flowers lacking bracteoles

39. Leafstalk (petiole) and sometimes rachis with 1 or more glands

40. Leaves usually with more than 6 pairs of leaflets, glands usually distributed between several pairs of leaflets, fertile stamens 10...........*Senna surattensis*

40. Leaves usually with 6 or fewer pairs of leaflets, glands only between the lower pair of leaflets, fertile stamens 6 or 7

41. Leaves with 2–3 pairs of leaflets, leaflets oblong-lanceolate.........................*Senna corymbosa*

41. Leaves with 3–5 pairs of leaflets, leaflets elliptic, obovate, or elliptic-lanceolate

42. Legume cylindric, 10–15 mm diameter, leaflets usually obovate or elliptic-obovate....
....................*Senna pendula* var. *glabrata*

42. Legume flattened, 5–7 mm broad, leaflets elliptic or elliptic-lanceloate....................
................*Senna mexicana* var. *chapmanii*

39. Leaf stalk lacking glands.............*Senna didymobotrya*

32. Plant a medium-size or large tree

43. Flowers greenish white, petals and sepals similar, usually borne in catkin-like racemes, trunk and larger branches often with large, stout, conspicuous, simple or branched thorns

44. Legume oblong, often twisted, usually 20–40 cm long with several seeds, leaf axis, petiole, and lower surface of leaflet distinctly finely hairy.......................*Gleditsia triacanthos*

44. Legume ovate, asymmetrical, 3–8 cm long; leaf axis, petiole, and lower surface of leaflet hairless or inconspicuously finely hairy with only a few hairs.............
....*Gleditsia aquatica*

43. Flowers pinkish or yellow, petals and sepals dissimilar, not borne in catkin-like racemes, trunk and larger branches lacking stout thorns

45. Petals 3, legume to about 12 cm long, more or less conspicuously constricted between the seeds.....................
....*Tamarindus indica*
45. Petals 5, legume usually at least 20 cm long, seed compartments moderately constricted between the seeds
  46. Leaves with 3–8 pairs of leaflets, flowers yellow...........
....*Cassia fistula*
  46. Leaves with 5–15 pairs of leaflets, flowers pink or reddish, fading to white....................*Cassia javanica*
31. Leaves odd pinnate, terminal leaflet or pinna present
  47. Leaflets alternate..*Dalbergia sisso*
  47. Leaflets opposite
    48. Corolla reduced to a single petal (the standard), flowers purple or pale lavender (sometimes nearly white), produced in an erect, narrow, terminal raceme
      49. Flowers purple, reddish purple, or violet; apex of leaflets notched or sometimes with the vein extending beyond the margin, but not appearing as a knoblike gland; leaflets usually entire.......................................*Amorpha fruticosa*
      49. Flowers white; apex of the leaflets usually terminated by a sessile, knoblike gland; leaflet margins distinctly or inconspicuously crenulate
        50. Margins of leaflets conspicuously crenulate, stalk of leaflet 1.5–2 mm long...*Amorpha herbacea* var. *crenulata*
        50. Margins of leaflets entire or inconspicuously crenulate, stalk of leaflet 1–1.5 mm long................................
.........................*Amorpha herbacea* var. *herbacea*
    48. Flowers not as above
      51. Stamens fused at the base or along a central axis (diadelphous or monadelphous), legume not necklace-like, not conspicuously constricted between the seed cavities
        52. Legume dry, several-seeded, with 4 papery wings...........
....*Piscidia piscipula*
        52. Legume not as above
          53. Calyx conspicuously lobed, fruit splitting at maturity
            54. Flowers pink to purple (rarely white), corolla usually 2–2.5 cm long, flower and fruit stalks slightly or abundantly glandular and hairy...........
....*Robinia hispida*
            54. Flowers white, corolla usually 1.5–2 cm long, flower and fruit stalks lacking glands and hairs.....
...............................*Robinia pseudoacacia*
          53. Calyx not lobed or merely toothed
            55. Inflorescence a raceme.............*Gliricidia sepium*

55. Inflorescence a pseudoraceme or panicle
  56. Legume elliptic or half-elliptic, 1-seeded........
    ....*Pongamia pinnata*
  56. Legume circular or oblong, 1 or more seeded..
    ....*Lonchocarpus punctatus*
51. Stamens free, not fused at base or along a central axis,
  legume necklace-like, swollen at intervals, conspicuously
  constricted between the seed cavities
  57. Leaflets of mature leaves densely hairy.....................
    ....*Sophora tomentosa* var. *occidentalis*
  57. Leaflets of mature leaves sparsely hairy....................
    ....*Sophora tomentosa* var. *truncata*
30. Leaves bipinnate
  58. Leaves with only 2 primary segments (pinna)
    59. Flowers bright pink, leaves with 7–10 pairs of elliptic to oblong-
      lanceolate or somewhat sickle-shaped leaflets 0.5–4.7 cm long.......
      ....*Calliandra haematocephala*
    59. Flowers bright yellow, leaves with numerous pairs of small leaflets
      2–4 mm long, often dropping early to leave a green, naked rachis....
      ....*Parkinsonia aculeata*
  58. Leaves with more than 2 primary segments
    60. Flowers greenish white, petals and sepals similar, usually borne in
      catkin-like racemes, trunk and larger branches often with large,
      stout, conspicuous, simple or branched thorns, leaves of two types
      (pinnate and bipinnate)
      61. Legume oblong, often twisted, usually 20–40 cm long with
        several seeds; leaf axis, petiole, and lower surface of leaflet
        distinctly finely hairy.........................*Gleditsia triacanthos*
      61. Legume, ovate, asymmetrical, 3–8 cm long; leaf axis, petiole,
        and lower surface of leaflet hairless or inconspicuously finely
        hairy with only a few hairs.......................*Gleditsia aquatica*
    60. Flowers otherwise, trunk lacking stout thorns, all leaves bipinnate
      62. Flowers individually inconspicuous, congregated into more or
        less congested heads or spikes, the corolla not immediately
        evident
        63. Stamens more than 10, usually numerous
          64. Stamens free, not joined at base into a sheath
            65. Flowers in an elongated spike
            ....*Acacia cornigera*
            65. Flowers in a rounded head
              66. Leaves usually with 10–25 pairs of primary
              segments (pinnae)..............*Acacia macracantha*
              66. Leaves with up to 8 pairs of primary segments
                67. Petiole gland more or less elliptic, spines

often fused at the base

  68. Spines massive, resembling miniature cow horns, pod inflated, long-pointed, 4–6 cm long, somewhat resembling the spines...............*Acacia sphaerocephala*

  68. Spines thin, pod linear, 4–13 cm long......
.................................*Acacia tortuosa*

67. Petiole gland circular, spines not fused at base

  69. Leaflets 1–2 cm long...........................
...........................*Acacia choriophylla*

  69. Leaflets less than 1 cm long

    70. Leaflets about 2 mm long, lacking secondary venation beneath..............
........................*Acacia pinetorum*

    70. Leaflets 4–6 mm long, with secondary venation beneath......*Acacia farnesiana*

64. Stamens joined at base into a sheath

  71. Leaves lacking glands on the petiole and rachis, legume splitting from the apex at maturity and recurving.................*Calliandra haematocephala*

  71. Leaves with glands on the petiole and/or rachis, fruit not as above

    72. Leaves with 8–12 primary segments (pinnae), fruit woody, in the shape of a flattened, rounded coil.................*Enterolobium contortisiliquum*

    72. Leaves with 1–4 pairs of primary segments, fruit not as above

      73. Stipules conspicuous on young growth

        74. Leaves with 10–30 pairs of leaflets, calyx hairy.................*Lysiloma latisiliquum*

        74. Leaves with 3–7 pairs of leaflets, calyx glabrous.....................*Lysiloma sabicu*

      73. Stipules spinelike or inconspicuous on young growth

        75. Petiole with a gland near the base, sometimes also between the primary leaf segments

          76. Flowers in compact yellow-white or greenish white heads to about 1.5 cm diameter

            77. Leaflets elliptic, to about 2.5 times longer than broad....................
......................*Albizia procera*

            77. Leaflets oblong, 3–5 times longer

than broad......*Albizia lebbekoides*

76. Flowers in rosy pink or creamy white
umbels 2.5–5 cm diameter

78. Umbels rosy pink, leaves with
5–15 pairs of primary segments....
..................…....*Albizia julibrissin*

78. Umbels creamy white, leaves with
2–5 pairs of primary segments…...
...............…..........*Albizia lebbeck*

75. Petiole lacking a gland near the base,
most glands positioned between the
primary leaf segments and between the
leaflets; leaves with 4 leaflets

79. Plants usually unarmed.................…....
.................…........*Pithecellobium keyense*

79. Plants usually spiny

80. Corolla 2.5–3 mm long...........…...
..…..............*Pithecellobium dulce*

80. Corolla 3–5 mm long

81. Leaflets thin, papery, mostly
longer than 2 cm...............…....
.....*Pithecellobium unguis-cati*

81. Leaflets thick, leathery, mostly
shorter than 2 cm...............…....
.....*Pithecellobium bahamense*

63. Stamens 10 or fewer

82. Plants armed with thorns at the leaf nodes, these usually
expressed as short, thorn-tipped axillary shoots............
.......................*Dichrostachys cinera* subsp. *africana*

82. Plants unarmed, or with prickles between the nodes

83. Leaves with 1 or more stalkless glands

84. Petiolar gland 1.5–2.5 mm long; margins of
leaflets ciliate; flower buds grayish hairy; stalk of
the legume 7–10 mm long, usually finely hairy....
......*Leucaena leucocephala* subsp. *leucocephala*

84. Petiolar glands 1–4 mm long; margins of the
leaflets glabrous; flower buds glabrous; stalk of
the legume 10–20 mm long, glabrous..............
..........…...*Leucaena leucocephala* subsp. *glabrata*

83. Leaves lacking glands

85. Plant a tree, flowers borne in slender racemes…...
.............................…...*Adenanthera pavonina*

85. Plant a low-growing, mat-forming vine or
sprawling, thicket-forming shrub, flowers borne

in rounded, congested heads
- 86. Stems and leaves armed with hooked prickles, fruit with 7–15 sgements...*Mimosa pigra*
- 86. Stems and leaves unarmed, fruit usually with 3 segments..................*Mimosa strigillosa*
- 62. Flowers individually conspicuous, not congregated into heads or spikes, the corolla evident
  - 87. Plant an erect, medium-sized tree
    - 88. Flowers red, petals 4–7 cm long, legume 20–60 cm long...*Delonix regia*
    - 88. Flowers yellow, petals to 3 cm long, legume 5–10 cm long
      - 89. Leaflets 1–2 cm long, corolla 3–4 cm long...*Peltophorum pterocarpum*
      - 89. Leaflets 0.5–1 cm long, corolla 2–3.5 cm long...*Peltophorum dubium*
  - 87. Plant an erect shrub (rarely a small tree), sometimes somewhat scrambling
    - 90. Petals 1.5–2 cm long, stamens to 5 cm long, protruding well beyond the corolla............*Caesalpinia pulcherrima*
    - 90. Petals less than 1 cm long, stamens to 1.5 cm long, protruding slightly from the corolla...*Caesalpinia pauciflora*

*◆**Abrus precatorius** Linnaeus. Rosary Pea, Blackeyed Susan. Disturbed sites, invading woodlands and hammocks. Central and southern peninsula, Marion, Volusia, and Hernando counties southward, including the Keys. Old World. FLEPPC listed (I).

*◆**Acacia auriculiformis** A. **Cunningham ex Bentham.** Earleaf Acacia. Disturbed sites, dunes, scrub, dry hammocks, margins of tidal marshes. Southern peninsula, especially along the coast, Brevard and Charlotte counties southward, including the Keys. Australia, New Guinea, Indonesia. FLEPPC listed (I).

!*Acacia choriophylla* Bentham. Cinnecord, Tamarindillo. Pine rocklands. Miami-Dade County and the Keys. State endangered.

*_Acacia cornigera_ (Linnaeus) Willdenow. Bullhorn Acacia. Disturbed sites, escaped from cultivation. Sparsely naturalized in the southwest peninsula, potentially from Pinellas to Lee counties and perhaps elsewhere. Central America.

Acacia farnesiana (Linnaeus) Willdenow. Sweet Acacia. Shell middens, coastal hammocks, pinelands. Sparsely distributed along the panhandle coast, more common in the central and southern peninsula from about Citrus, Marion, and Volusia counties south, including the Keys.

Acacia macracantha Humboldt & Bonpland ex Willdenow. Long Spine Acacia, Steel Acacia, Porknut. Margins of mangroves, hammocks, sandy ridges. Naturalized from cultivation in Manatee County, native but perhaps extirpated in Miami-Dade County and the Keys.

Acacia pinetorum **F. J. Hermann.** Pineland Acacia. Shell middens, coastal hammocks, pinelands. Mostly southern peninsula from Lee and Broward counties southward including the Keys, disjunct to Brevard, Citrus, Pasco, and Pinellas counties.

*Acacia retinodes **Schlechtendal.** Water Wattle. Disturbed sites. Sparsely naturalized. Potentially from Glades County to the Keys. Australia.

*Acacia sphaerocephala **Schlechtendal & Chamisso.** Bee Wattle. Subtropical hammocks. Southern peninsula. Collier to Miami-Dade counties; absent from the Keys. Mexico.

!Acacia tortuosa **(Linnaeus) Willdenow.** Poponax. Shell middens, disturbed sites. Collier County. State endangered.

*◆Adenanthera pavonina **Linnaeus.** Red Sandalwood, Red Beadtree. Disturbed sites. Southern peninsula, naturalized or potentially so from Lee County to Miami-Dade County and the Keys. Tropical Asia. FLEPPC listed (II).

*◆Albizia julibrissin **Durazzini.** Silktree, Mimosa. Disturbed sites, wetland edges, roadsides. Panhandle, east to St. Johns County, south to Hillsborough, Polk, and Osceola counties. Asia. FLEPPC listed (I).

*◆Albizia lebbeck **(Linnaeus) Bentham.** Woman's Tongue, Lebbeck's Albizia. Disturbed sites, yards, invading hammocks. Central and southern peninsula, Marion County south to the Keys. Asia. FLEPPC listed (I).

*Albizia lebbekoides **(de Candolle) Bentham.** Indian Albizia. Disturbed sites, scarcely naturalized. Miami-Dade County. Tropical Asia.

*Albizia procera **(Roxburgh) Bentham.** Tall Albizia. Disturbed sites, escaped from cultivation. Miami-Dade County. Asia to Australia.

Amorpha fruticosa **Linnaeus.** False Indigo, Bastard Indigo, Bastard False Indigo. Wet woodlands, stream and river margins, floodplain woodlands. Throughout except the Keys.

Amorpha herbacea **Walter var.** *herbacea.* Clusterspike False Indigo. Scrub, sandhills, flatwoods. Central panhandle west to Nassau County, south to Collier County, mostly in the western half of the peninsula.

!Amorpha herbacea **Walter var.** *crenulata* **(Rydberg) Isely.** Miami Lead Plant. Pine rocklands. Miami-Dade County. State endangered.

Apios americana **Medikus.** Groundnut. Hammocks, floodplain forests. Essentially throughout, except the Keys.

*Bauhinia aculeata **Linnaeus.** White Orchid Tree. Disturbed sites, escaped from cultivation. Miami-Dade County. Central and South America.

*Bauhinia purpurea **Linnaeus.** Purple Orchid Tree, Butterfly Tree. Disturbed sites, escaped from cultivation. Southeast peninsula, St. Lucie County to Miami-Dade County. Tropical Asia.

*◆Bauhinia variegata **Linnaeus.** Orchid Tree, Mountain Ebony. Disturbed sites, escaped from cultivation, invading pine rocklands, mesic flatwoods, hammocks, swamp margins. Southern peninsula, Brevard and Polk counties, south to Miami-Dade County. FLEPPC listed (I).

*Bauhinia yunnanensis **Franchet.** Yunnan Bauhinia. Disturbed sites, escaped from cultivation. Miami-Dade County. China.

Caesalpinia bonduc **(Linnaeus) Roxburgh.** Gray Nicker, Gray Nicker Bean, Sea Bean,

Fever Nut, Hold-Back. Coastal strand, margins of mangrove forests. East and west coasts from Levy and Volusia counties south to the Keys.

Caesalpinia crista **Linnaeus.** Yellow Nicker. Disturbed sites, mostly along the coast, escaped from cultivation. Southern peninsula, documented naturalized in Lee County. Malaysia.

!*Caesalpinia major* **(Medikus) Dandy & Exell.** Yellow Nicker, Yellow Nicker Bean, Hawaii Pearls. Coastal strand, margins of mangrove forests. Southeastern peninsula, Martin County south to the Keys. State endangered.

!*Caesalpinia pauciflora* **(Grisebach) C. Wright**. Fewflower Hold-Back, Caesalpinia. Subtropical hammocks, pinelands. Keys. State endangered.

Caesalpinia pulcherrima **(Linnaeus) Swartz.** Pride-of-Barbados, Dwarf Poinciana. Disturbed sites, escaped from cultivation. Southern peninsula, Lee and Miami-Dade counties south to the Keys. Asia.

Cajanus cajan **(Linnaeus) Huth.** Pigeonpea. Disturbed sites, escaped from cultivation. Southeast peninsula, Broward County south to the Keys. Old World.

Calliandra haematocephala **Hasskarl.** Powderpuff Tree. Sparingly and irregularly naturalized from cultivation. Escambia, Brevard, and Broward counties. Tropical America.

Canavalia rosea **(Swartz) de Candolle.** Baybean, Seaside Bean, Seaside Jackbean. Coastal strand. Coastal counties, Dixie and Volusia counties south to the Keys.

Cassia fistula **Linnaeus.** Golden Shower. Disturbed sites, escaped from cultivation. Miami-Dade County, presumably elsewhere. Tropical Asia.

Cassia javanica **Linnaeus var.** *indochinensis* **Gagnepain.** Pink Shower. Disturbed sites, more likely only cultivated and not naturalized. Miami-Dade County. Tropical Asia.

Cercis canadensis **Linnaeus.** Eastern Redbud. Rich moist woods, roadsides, yards, upland forests. Panhandle and northern peninsula, east to Clay County, south to Polk County.

!*Chamaecrista lineata* **(Swartz) Greene var.** *keyensis* **(Pennell) Irwin & Barneby.** Big Pine Partridge Pea, Key Cassia, Narrowpod Sensitive Pea. Pinelands, margins of subtropical hammocks. Miami-Dade County and the Keys. State endangered.

!*Dalbergia brownei* **(Jacquin) Schinz.** Coinvine. Browne's Indian Rosewood. Mangroves, margins of subtropical hammocks. Miami-Dade County and the Keys. State endangered.

Dalbergia ecastophyllum **(Linnaeus) Taubert.** Coinvine. Shell mounds within mangrove wetlands, coastal hammocks, margins of mangrove forests, coastal strand. Coastal and near coastal counties of the southern peninsula, Pinellas and Brevard counties south to the Keys.

*◆*Dalbergia sissoo* **Roxburgh ex de Candolle.** Indian Rosewood. Disturbed sites, escaped from cultivation. Southwestern and southeastern peninsula, Pinellas and Hillsborough counties south along the coast to Collier County; Broward County south to the Keys. India. FLEPPC listed (II).

Delonix regia **(Bojer ex Hooker) Rafinesque.** Royal Poinciana. Disturbed sites, streetsides, escaped from cultivation. Southern peninsula, Lee and Broward counties southward, including the Keys. Madagascar.

Dichrostachys cinerea **(Linnaeus) Wight & Arnott subsp.** *africana* **Brenan & Brum-**

mitt. Aroma. Disturbed sites. Irregularly and sparingly naturalized, central peninsula, southeast peninsula, Keys. Old World.

*Enterolobium contortisiliquum **(Vellozo) Morong.** Earpod Tree. Disturbed sites, escaped from cultivation. Irregularly naturalized, central to south-central peninsula, Pinellas, Lake, and Volusia counties south to Broward County. Tropical America.

Erythrina herbacea **Linnaeus.** Coralbean, Cardinal Spear, Cherokee Bean. Pinelands, hammocks, dry uplands. Essentially throughout, including the Keys.

Gleditsia aquatica **Marshall.** Water Locust. Floodplains, river swamps, low hammocks. Panhandle, south to Sarasota, DeSoto, and Brevard counties.

Gleditsia triacanthos **Linnaeus.** Honey Locust. Moist hammocks, well-drained woods, margins of depressions in upland woodlands. Panhandle east to Levy County.

Gleditsia × texana **Sargent.** Texas Honey Locust. Moist coastal woodlands bordering streams and rivers. Central panhandle, Jefferson and Wakulla counties and perhaps elsewhere.

*Gliricidia sepium **(Jacquin) Kunth ex Walpers.** Quickstick. Disturbed sites. Keys. Mexico, Central America, perhaps South America.

*Lespedeza bicolor **Turczaninow.** Shrubby Lespedeza. Thunberg's Disturbed sites, fields, roadsides, escaped from cultivation. Essentially the panhandle, also reported naturalized in Henando County. Japan.

*Lespedeza thunbergii **(de Candolle) Nakai.** Thunberg's Lespedeza. Disturbed sites, roadsides, fallow fields. Sparsely naturalized in the panhandle, Escambia to Jefferson counties. Asia.

*◆Leucaena leucocephala **(Lamarck) de Wit.** Lead Tree, Jumbie Bean. Disturbed sites, hammocks, coastal strand. Alachua County (where not common) and Hillsborough County south to the Keys, predominately in the more southern parts of the state. West Indies. FLEPPC listed (II).

*Leucaena leucocephala **(Lam.) de Wit subsp.** *glabrata* **(Rose) Zárate.** Lead Tree, White Lead Tree. Disturbed sites, escaped from cultivation. Pinellas and Volusia counties southward. Mexico, Central America.

*Lonchocarpus punctatus **Kunth.** Dotted Lancepod. Subtropical hammocks, disturbed sites, escaped from cultivation. Keys. Tropical South America.

!Lupinus westianus **Small var.** *westianus.* Gulf Coast Lupine. Dunes, sandhills. Western panhandle, Escambia to Franklin counties. State threatened.

!Lupinus westianus **Small var.** *aridorum* **(McFarlin ex Beckner) Isely.** Beckner's Lupine, McFarlin's Lupine. Scrub. Central peninsula; Orange, Polk, and Osceola counties. State and federally endangered.

Lysiloma latisiliquum **(Linnaeus) Bentham.** False Tamarind. Pinelands, hammocks, pine—palmetto woodlands, disturbed sites, mostly near the coast. Southern peninsula, Collier and Miami-Dade counties to the Keys.

*Lysiloma sabicu **Bentham.** Horseflesh Mahogany. Hammocks, disturbed sites. Miami-Dade County. West Indies (especially Cuba and Hispaniola).

*◆Mimosa pigra **Linnaeus.** Black Mimosa. Disturbed wetlands. South-central peninsula, Highlands and Okeechobee counties south to Broward County. Tropical America. FLEPPC listed (I).

Mimosa strigillosa **Torrey & A. Gray.** Sunshine Mimosa, Powderpuff. Sunny disturbed sites, cultivated along roadsides. Leon to Miami-Dade counties.

*__Mucuna pruriens__ **(Linnaeus) de Candolle.** Cowitch, Velvetbean. Disturbed sites, escaped from cultivation. Southern peninsula, Lee and Palm Beach counties southward, including the Keys. Asia.

*__Mucuna sloanei__ **Fawcett & Rendle.** Horse-eye Bean. Disturbed sites, escaped from cultivation. Broward and Miami-Dade counties. West Indies, Central and South America.

*__Parkinsonia aculeata__ **Linnaeus.** Jerusalem Thorn, Mexican Palo Verde. Disturbed sites, roadsides. Naturalized at scattered sites throughout the state. Mexico, tropical America, southwestern United States.

*__Peltophorum dubium__ **(Sprengel) Taubert.** Horsebush. Disturbed sites, escaped from cultivation. Naturalized at scattered sites in the central and southern peninsula; Seminole, Orange, and Miami-Dade counties. South America.

*__Peltophorum pterocarpum__ **(de Candolle) Baker ex K. Heyne.** Yellow Poinciana. Disturbed sites, escaped from cultivation. Sparingly naturalized in the southern peninsula, Lee County to Miami-Dade County, Keys. Sri Lanka, Malay archipelago, Indonesia, northern Australia.

__Piscidia piscipula__ **(Linnaeus) Sargent.** Florida Fishpoison Tree, Jamaica Dogwood, Fishfuddle Tree. Coastal strand, subtropical hammocks, shell middens, pine rocklands. Pinellas and Hillsborough counties south to the Keys.

__Pithecellobium bahamense__ **Northrop.** Bahama Blackbead. Pine rocklands. Keys.

*__Pithecellobium dulce__ **(Roxburgh) Bentham.** Monkeypod. Disturbed sites, escaped from cultivation. Southeastern and southern peninsula, Manatee to Miami-Dade counties. Tropical America.

!__Pithecellobium keyense__ **Britton ex Britton & Rose.** Florida Keys Blackbead. Coastal hammocks. Southeastern peninsula, Martin County south to Monroe and Miami-Dade counties, including the Keys. State threatened.

__Pithecellobium unguis-cati__ **(Linnaeus) Bentham.** Catclaw Blackbead. Coastal hammocks, shell mounds and middens. Brevard County, southeastern peninsula, Hillsborough to Miami-Dade counties and the Keys.

*__Pongamia pinnata__ **(Linnaeus) Pierre.** Pongam, Karum Tree, Poonga-oil Tree. Disturbed sites, margins of hammocks, escaped from cultivation. Southwestern and southeastern peninsula, Sarasota, Palm Beach, Broward counties, perhaps elsewhere. Southeast Asia.

*◆__Pueraria montana__ **(Loureiro) Merrill var.** _lobata_ **(Willdenow) Maesen & S. M. Almeida ex Sanjappa & Predeep.** Kudzu. Disturbed sites, margins of pinelands and upland hammocks. Nearly throughout, more common in the northern peninsula and panhandle. Asia. FLEPPC listed (I).

*__Robinia hispida__ **Linnaeus.** Bristly Locust. Sandhills, dry upland woods, disturbed sites, presumably escaped from cultivation. Sparingly naturalized in scattered locations across the panhandle and northwestern peninsula, Escambia to Alachua counties. Lower mountains and Piedmont, southeastern United States.

*__Robinia pseudoacacia__ **Linnaeus.** Black Locust. Upland woods, disturbed sites, roadsides, margins of pastures, fence lines. Irregularly naturalized, panhandle, east and south to Clay, Marion, and Lake counties. Southeastern United States, originally mostly in the

Southern Appalachians, now widespread across the East.

*_Senna corymbosa_ (Lamarck) Irwin & Barneby. Argentine Wild Sensitive Plant. Disturbed sites, escaped from cultivation. Miami-Dade County. South America.

*_Senna didymobotrya_ (Fresenius) Irwin & Barneby. African Wild Sensitive Plant. Disturbed sites, escaped from cultivation. Miami-Dade County. Tropical Africa.

!_Senna mexicana_ (Jacquin) H. S. Irwin & Barneby var. _chapmanii_ (Isely) Irwin & Barneby. Chapman's Wild Sensitive Plant, Bahama Senna. Pinelands, roadsides, hammocks, dunes. Miami-Dade County and the Keys. State threatened.

*◆_Senna pendula_ (Humbboldt & Bonpland ex Willdenow) H. S. Irwin & Barneby var. _glabrata_ (Vogel) Irwin & Barneby. Climbing Cassia, Christmas Cassia, Christmas Senna, Valamuerto. Disturbed sites, mesic and hardwood hammocks, basin swamps, escaped from cultivation. Central and southern peninsula, Pinellas, Hillsborough, Osceola, and Brevard counties southward, including the Keys. South America. FLEPPC listed (I).

*_Senna surattensis_ (Burman f.) Irwin & Barneby. Glossy Shower. Disturbed sites, escaped from cultivation. Miami-Dade County, the Keys, and presumably elsewhere. Old World tropics.

Sesbania drummondii (Rydberg) Cory. Poisonbean. Disturbed sites, margins of freshwater and brackish marshes. Western panhandle, Escambia, and Santa Rosa counties.

*_Sesbania grandiflora_ (Linnaeus) Persoon. Vegetable Hummingbird. Disturbed sites, escaped from cultivation. Keys. Tropical Asia, including India, Indonesia, Malaysia, Myanmar, Phillipines.

Sesbania herbacea (Miller) McVaugh. Danglepod. Disturbed sites, roadsides, wet ditches. Essentially throughout.

*◆_Sesbania punicea_ (Cavanilles) Bentham. Rattlebox, Purple Sesban. Disturbed sites, wetland margins, margins of tidal marshes, moist roadsides. Panhandle south to Lee, Polk, and Indian River counties. South America. FLEPPC listed (II).

*_Sesbania sericea_ (Willdenow) Link. Silky Sesban. Disturbed sites. Broward County and the Keys. Africa.

*_Sesbania virgata_ (Cavanilles) Persoon. Wand Riverhemp. Margins of coastal marshes. Sparingly naturalized, Escambia, Santa Rosa, Gulf, and Pinellas counties, perhaps elsewhere. South America.

*_Sophora tomentosa_ Linnaeus var. _occidentalis_ (Linnaeus) Isely. Yellow Necklacepod. Disturbed sites. Southern peninsula. Sparingly naturalized, Sarasota, Martin, and Miami-Dade counties, potentially elsewhere. Texas, tropical America.

Sophora tomentosa Linnaeus var. _truncata_ Torrey & A. Gray. Yellow Necklacepod. Hammocks, coastal woodlands. Coastal counties, Levy and Brevard counties south to the Keys.

*_Tamarindus indica_ Linnaeus. Tamarind. Disturbed sites, coastal hammocks. Southern peninsula; irregularly naturalized from Manatee and Broward counties southward to the Keys. Old World tropics, especially India.

*_Wisteria floribunda_ (Willdenow) de Candolle. Japanese Wisteria. Disturbed sites, escaped from cultivation. Western panhandle, north-central peninsula. Japan.

Wisteria frutescens (Linnaeus) Poiret. American Wisteria. Margins of bay swamps and wetland drainages, wet hammocks, river swamps. Panhandle, Escambia County east to Clay and Nassau counties, potentially south to Orange County. China.

*◆*Wisteria sinensis* (Sims) **Sweet.** Chinese Wisteria. Disturbed sites, roadsides, wood-land margins, escaped from cultivation. Panhandle east to Clay and Duval counties, south to Brevard County. China. FLEPPC listed (II).

*Wisteria × formosa **Rehder.** Plants that do not effectively key to either *W. floribunda* or *W. sinensis*, especially in leaflet number, may represent this presumed hybrid of our two Asian species.

Fagaceae (Oak or Beech Family)

1. Buds clustered at the tip of the twig, with a true terminal bud closely subtended by several lateral buds, fruit an acorn (the oaks, genus *Quercus*)
  2. Leaves coarsely toothed, with few to numerous, blunt to sharply pointed teeth, each conspicuously or obscurely terminated by a lateral vein (white oaks, in part, section Quercus)
    3. Plant a tree
      4. Mature nut averaging 15–20 mm long, leaves glabrate or with scattered pubescence beneath, secondary leaf veins on the majority of leaves numbering less than 13 per side, bark gray, flaky, plants typically occurring in calcareous situations, usually in association with limestone…...................... ....…...*Quercus muehlenbergii*
      4. Mature nut averaging 25–35 mm long, leaves felty pubescent beneath, secondary leaf veins on the majority of leaves numbering 13 or more per side, bark whitish, scaly, plants usually occurring on mesic slopes or in more or less moist bottoms...*Quercus michauxii*
    3. Plant a low, clonal shrub...*Quercus minima*
  2. Most or all leaves entire or terminally lobed, or pinnately lobed or sinuate, not distinctly toothed (white and red oaks)
    5. Most leaves entire or only terminally lobed, not distinctly pinnately lobed or sinuate
      6. Some or many mature leaves on any tree spatulate in outline
        7. Leaves often varying widely in shape, with at least 5 shapes often present, lower surface of leaf glabrous or with at most tufts of hairs in the vein axils... ....*Quercus nigra*
        7. Leaves more or less uniform in size, lower surface of leaf densely matted with grayish, appressed, stellate pubescent (requires 20× magnification to see clearly)…...*Quercus durandii*
      6. Leaves varying in shape, but not spatulate in outline
        8. Petiole glabrous
          9. Leaf blade more or less smooth above, essentially flat, margins slightly

revolute, lower leaf surface with tufts of hairs in the veins axils.................
....*Quercus myrtifolia*

 9. Leaf blade more or less rugose above, convex with revolute margins,
  lower leaf surface typically with yellowish, scurfy pubescence, central
  vein with scattered hairs, vein axils lacking tufts of hairs..*Quercus inopina*

8. Petiole pubescent

 10. Blade convex above, margins revolute, sometimes strongly so

  11. Upper surface of leaf smooth, lower surface more or less glabrous
   except for tufts of hairs in the main vein axils........*Quercus myrtifolia*

  11. Upper surface of leaf rugose, veins sometimes deeply impressed,
   lower surface densely coated with minute, grayish white hairs and
   somewhat felty to the touch....................................*Quercus geminata*

 10. Blade flat, margins not conspicuously revolute

  12. Lower leaf surface densely covered with grayish, matted pubescence
   that obscures the blade tissue…...............................*Quercus virginiana*

  12. Lower leaf surface lacking a dense covering of grayish, matted
   pubescence

   13. Leaves varying broadly obovate, broadly oblanceolate, or broadly
    rhombic, not greater than 2–3 times longer than broad

    14. At least some leaves with marginal bristles or shallow, bristle-
     tipped lobes

     15. Blades longer than wide, lower surface with tufts of hairs in
      the vein axils..*Quercus arkasana*

     15. Blades about as long as wide, lower surface with scattered
      pubescence, but lacking tufts of hairs in the veins
      axils..*Quercus marilandica*

    14. Margins and marginal lobes, if present, lacking bristles…..........
    ....*Quercus chapmanii*

   13. Leaves narrowly elliptic, narrowly obovate, or narrowly
    oblanceolate, averaging more than 3 times longer than wide

    16. Petiole glabrous

     17. Lower surface of leaf glabrous........*Quercus hemisphaerica*

     17. Lower surface of leaf with tufts of hairs in some or all of
      the main vein axils

      18. Leaves linear or narrowly elliptic..........*Quercus phellos*

      18. Leaves broadly elliptic or rhombic, at least some leaves
       on any tree conspicuously widest near the middle...........
       ....*Quercus laurifolia*

    16. Petiole pubescent

     19. Upper surface of leaf bluish green with raised veins, lower
      surface with tufts of reddish hairs in the vein axils, plant
      typically a small tree....................................*Quercus incana*

19. Upper surface of leaf green with impressed veins, lower surface uniformly pubescent with grayish or grayish brown hairs, lacking tufts in the vein axils, plant typically a low, clonal shrub..*Quercus pumila*

5. Most or all leaves pinnately lobed or sinuate (white or red oaks)

  20. Lobe apices bristle tipped (red oaks, section Lobatae)

    21. Lower surface of leaf pubescent

      22. Pubescence of lower surface of leaves thinly and finely distributed throughout the surface, at least on young leaves, or confined to major veins or vein axils, surface tissue of lower surface easily visible despite the pubescence..*Quercus velutina*

      22. Pubescence of lower surface dense and matted

        23. Base of blade U-shaped, basal margins of blade meeting the petiole opposite one another or often slightly offset, terminal lobe often much longer than lateral lobes, pubescence on lower surface often rusty..*Quercus falcata*

        23. Base of blade angled, truncate, or short tapered, not distinctly U-shaped, pubescence of lower surface gray...............*Quercus pagoda*

    21. Lower surface of leaf glabrous except for tufts of hairs in the vein axils and sometimes along major veins

      24. Petiole to about 1.5 cm long...*Quercus laevis*

      24. Petiole 2 cm long or longer

        25. Petiole often greater than 1/3 the length of the blade, terminal buds pubescent, bark dark gray to nearly black and coarsely furrowed...*Quercus velutina*

        25. Petiole often less than 1/3 the length of the blade, terminal buds glabrous, bark gray, more or less smooth, not deeply and coarsely furrowed..*Quercus shumardii*

  20. Lobe apices rounded or sharply angled, not bristle-tipped (white oaks, in part, section Quercus)

    26. Lobes of mature leaves sharply angled or acute at apex, cup of acorn covering and nearly concealing the nut...............................*Quercus lyrata*

    26. Lobes of mature leaves rounded or squarish at apex, or leaves sparingly lobed, cup of the acorn concealing less than 1/2 of the nut

      27. Mature leaf blades appearing glabrous beneath, at least to the naked eye, hairs, if present, minute and inconspicuous, twigs glabrous

        28. Mature crown leaves deeply and more or less evenly lobed, lobing of leaves of immature trees or sometimes the lower branches of mature trees at least regularly lobed and with leaves of a uniform size, mature leaves of crown branches often with 7 or more lobes, the sinuses between the lobes often deeper than 1/2 the length of the adjacent lobes.......................................*Quercus alba*

        28. Leaves shallowly and irregularly lobed, lobes typically 7 or fewer

29. Leaf sinuate, entire and obovate, or with 3 mostly terminal lobes, lower surface of leaf densely covered with minute, silvery, appressed stellate pubescence that requires magnification to see clearly..............................*Quercus durandii*
29. Leaves with up to about 5 shallow lobes, lower surface with minute, erect, branched pubescence.................*Quercus austrina*
27. Mature blades pubescent beneath, pubescence usually easily visible to the naked eye or with 10× magnification
30. Twigs of the season densely stellate pubescent with tawny hairs..*Quercus stellata*
30. Twigs of the season glabrous or sparsely hairy with whitish, erect, branched hairs
31. Lower surfaces of leaves stellate pubescent with whitish, erect hairs...*Quercus margaretta*
31. Lower surfaces of leaves stellate pubescent with appressed, matted hairs.............…...........................*Quercus durandii*
1. Buds not so disposed, fruit consisting of several seeds contained within a spiny involucre
32. Buds elongate, 1–2 cm long, resembling a cigar or torpedo, bark very smooth, not at all ridged or furrowed, inflorescence a ball-like, dangling head.....................
....….......*Fagus grandifolia*
32. Buds egg-shaped or oblong, 2–4 mm long, inflorescence slender, spikelike, elongate
33. Lower surface of leaf glabrous or finely hairy
34. Leaf margin coarsely toothed, the lateral veins extending beyond the tip of the teeth and forming a bristle...........…......................*Castanea dentata*
34. Leaf margin bluntly toothed, the terminus of the lateral veins extending slightly beyond the tip of the teeth, forming a blunt mucro......................
....…..............*Castanea crenata*
33. Lower surface of leaf distinctly and often densly hairy
35. Mature leaves mostly longer than 13 cm......................*Castanea mollissima*
35. Mature leaves mostly 13 cm long or shorter................…....*Castanea pumila*

Castanea crenata **Siebold & Zuccarini.** Japanese Chestnut. Roadsides, pinelands, disturbed sites, escaped from cultivation. Rarely if at all naturalized. Reported from Marion County. Japan, South Korea.

Castanea dentata **(Marshall) Borkhausen.** American Chestnut. Dry uplands, mesic hammocks. Western panhandle, Escambia and Okaloosa counties. Once represented in Florida by mature trees, likely represented today only by root sprouts of diseased trees.

Castanea mollissima **Blume.** Chinese Chestnut. Disturbed sites, pinelands, roadsides. Sparingly naturalized or under-reported. Central panhandle, nothern peninsula. China, Korea.

Castanea pumila **(Linnaeus) Miller.** Chinquapin. Pinelands, dry bluffs and slopes, hammocks, sandhills. Panhandle, Escambia County east to Nassau County, south to Hillsborough, Lake, and Orange counties.

Fagus grandifolia **Ehrhart.** American Beech. Rich slope forests, bluffs, mesic hammocks, often in association with limestone. Panhandle, Escambia County east to Hamilton, Columbia, and Alachua counties.

Quercus alba **Linnaeus.** White Oak. Rich slope forests, mesic hammocks, temperate hardwood forests, bluffs. Panhandle, Okaloosa to Taylor and Suwannee counties.

!*Quercus arkansana* **Sargent.** Arkansas Oak. Well-drained, sandy pine—oak—hickory woodlands, dry hammocks. Central panhandle, Santa Rosa to Walton counties; reported but questionable from Calhoun County. State threatened.

Quercus austrina **Small.** Bastard White Oak. Bluffs, upland forests, slopes, ravines. Panhandle and northern peninsula, Escambia to Nassau counties, south to Citrus, Lake, and Volusia counties.

Quercus chapmanii **Sargent.** Chapman's Oak. Sandy uplands, sandhills, sand pine—oak scrub, scrubby flatwoods. Coastal counties in the panhandle, east to Nassau County, south to Miami-Dade County.

Quercus durandii **Buckley.** Durand's Oak, Basket Oak, Bastard White Oak. Dry or moist hammocks, often in association with limestone. Erroneously reported for the panhandle, perhaps not present in Florida but potentially so.

Quercus falcata **Michaux.** Spanish Oak, Southern Red Oak. Upland woods, sandhills, dry bluffs. Panhandle and northern peninsula, Escambia County east to Duval County, south to Polk County.

Quercus geminata **Small.** Sand Live Oak. Sandhills, coastal hammocks, pine—oak scrub, inland dunes. Nearly throughout, panhandle and peninsula, south to Palm Beach and Collier counties.

Quercus hemisphaerica **W. Bartram ex Willdenow.** Darlington Oak. Mixed woodlands, disturbed uplands, sand ridges, upper reaches of dry slopes. Panhandle and peninsula, south to about Palm Beach County.

Quercus incana **W. Bartram.** Bluejack Oak. Sandhills, xeric uplands. Panhandle and northern peninsula, south to Sarasota, DeSoto, and Highlands counties.

Quercus inopina **Ashe.** Scrub Oak. Sand pine scrub, flatwoods. Central peninsula, St. Johns and Putnam south to Glades and Martin counties.

Quercus laevis **Walter.** Turkey Oak. Sandhills, dry flatwoods, well-drained sandy ridges. Panhandle and northern peninsula, south to Martin and Collier counties.

Quercus laurifolia **Michaux.** Laurel Oak, Swamp Laurel Oak, Diamond-leaf Oak. Floodplains, wet hammocks, bottomlands. Panhandle and northern peninsula, south to the southern peninsula; absent from the Keys.

Quercus lyrata **Walter.** Overcup Oak. Floodplains and adjacent lower slopes of large rivers. Panhandle, east to Columbia, Union, Alachua, and Levy counties.

Quercus margaretta **Ashe ex Small.** Sand Post Oak. Sandhills, coastal sandy ridges, pine—oak scrub. Panhandle and northern peninsula, south to Hillsborough and Polk counties.

Quercus marilandica **Münchhausen.** Blackjack Oak. Dry upland woods and pinelands, often where clay is present. Panhandle, Jefferson County westward.

Quercus michauxii **Nuttall.** Basket Oak, Swamp Chestnut Oak. Moist hammocks, rich slope and ravines forests, margins of the highest portions of floodplains, often where limestone is present. Panhandle, east to Duval County, south to Hernando and Lake counties.

Quercus minima **(Sargent) Small.** Dwarf Live Oak. Sandhills, flatwoods. Essentially throughout, except the Keys.

Quercus muehlenbergii **Engelmann.** Chinquapin Oak. Upland hardwood forests, mesic hammocks, usually in association with limestone. Central panhandle, Walton to Jefferson counties.

Quercus myrtifolia **Willdenow.** Myrtle Oak. Scrub, dunes, flatwoods, deep inland and coastal sands. Western and coastal panhandle; peninsula from Duval and Levy counties southward; absent from the Keys.

Quercus nigra **Linnaeus.** Water Oak. Disturbed uplands, floodplains, hammocks, streambanks, flatwoods. Panhandle and northern peninsula, south to Collier and Broward counties.

Quercus pagoda **Rafinesque.** Cherrybark Oak, Swamp Red Oak. Upland woods, bottomlands, floodplains, moist slopes adjacent to floodplains. Central panhandle, mostly in association with the Apalachicola River.

Quercus phellos **Linnaeus.** Willow Oak. Floodplains, bottomlands, disturbed uplands. Sparsely distributed from Escambia to Taylor, Columbia, and Duval counties.

Quercus pumila **Walter.** Running Oak. Sandhills, flatwoods. Panhandle, east to Duval County, south to Miami-Dade County.

Quercus shumardii **Buckley.** Shumard Oak. Rich slope forests, moist upland woods, riverbanks, moist hammocks, often in association with limestone. Central and eastern panhandle and north-central peninsula.

Quercus stellata **Wangenheim.** Post Oak. Sandhill pinelands, upland mixed woods, often where clay or limestone is present. Panhandle, Escambia to Jefferson counties; central ridge of the northern peninsula, Columbia to Marion counties.

Quercus velutina **Lamarck.** Black Oak. Mixed upland woods, sandhill pinelands, often where clay is present. Confined largely to the central panhandle, Holmes to Leon and Wakulla counties; Suwannee County.

Quercus virginiana **Miller.** Live Oak. Disturbed uplands, moist or wet hammocks, coastal hammocks, flatwoods, margins of coastal marshes, pastures. Throughout, including the Keys.

Recognized Named Hybrids

Quercus × *asheana* Little. (*Q. incana* ×*Q. laevis*)
Quercus × *atlantica* Ashe. (*Q. incana* × *Q. laurifolia*)
Quercus × *caduca* Trelease. (*Q. incana* × *Q. nigra*)
Quercus × *comptoniae* Sargent. (*Q. lyrata* × *Q. virginiana*)
Quercus × *harbisonii* Sargent. (*Q. geminata* × *Q. margaretta*)
Quercus × *mellichampii* Trelease. (*Q. hemisphaerica* × *Q. laevis*)
Quercus × *neopalmeri* Sudworth ex E. J. Palmer. (*Q. nigra* × *Q. shumardii*)
Quercus × *oviedoensis* Sargent. (*Q. myrtifolia* × *Q. incana*)
Quercus × *rolfsii* Small. (*Q. chapmanii* × *Q. minima*)

Quercus × *rudkinii* **Britton.** (*Q. marilandica* × *Q. phellos*)
Quercus × *subintegra* (**Engelmann**) **Trelease.** (*Q. falcata* × *Q. incana*)
Quercus × *succulenta* **Small.** (*Q. geminata* × *Q. minima*)
Quercus × *venulosa* **Ashe.** (*Q. arkansana* × *Q. incana*)
Quercus × *walteriana* **Ashe.** (*Q. laevis* × *Q. nigra*)

Gelsemiaceae (Gelsemium Family)

1. Sepals blunt or rounded at the apex; flowering mid-winter or very early spring...........
....*Gelsemium sempervirens*
1. Sepals acuminate at apex; flowering mid-spring, near and beyond the end of the typical
flowering time of *G. sempervirens*...*Gelsemium rankinii*

***Gelsemium rankinii* Small.** Rankin's Jessamine, Swamp Jessamine. Bogs, swamps,
floodplains, ravine bottom streams. Panhandle, Escambia to Leon and Wakulla counties,
disjunct to Hamilton and Nassau counties.
***Gelsemium sempervirens* (Linnaeus) Aiton f.** Carolina Jessamine, Yellow Jessamine,
Evening Trumpetflower, Poor Man's Rope. Upland woods, hammocks. Panhandle and
peninsula, south to Charlotte and Palm Beach counties.

Goodeniaceae (Goodenia Family)

1. Central flower of inflorescence sessile, sepals less than 2 mm long, fruit black...............
....*Scaevola plumiera*
1. Central flower of inflorescence pedicellate, sepals greater than 2 mm long, fruit white
or yellowish white
    2. Leaves glabrous or nearly so..*Scaevola taccada* var. *taccada*
    2. Leaves with appressed, silky hairs...............................*Scaevola taccada* var. *sericea*

!*Scaevola plumieri* (**Linnaeus**) **Vahl.** Scaevola, Beachberry, Inkberry, Gullfeed. Coastal
strand. Southern peninsula, mostly along the coast from Pinellas, Hillsborough, and Brevard counties southward. State threatened.
*◆*Scaevola taccada* (**Gaertner**) **Roxburgh var. *taccada*.** Beach Naupaka. Coastal
strand. Sarasota and Indian River counties southward, including the keys. Coastlines of
the Indian and western Pacific Oceans. FLEPPC listed (I).
*◆*Scaevola taccada* (**Gaertner**) **Roxburgh var. *sericea* (Vahl) H. St. John.** Beach
Naupaka. Coastal strand. Pinellas and Brevard counties southward, including the Keys.
Coastlines of the Indian and western Pacific Oceans. FLEPPC listed (I).

Grossulariaceae (Gooseberry Family)

!*Ribes echinellum* (Coville) **Rehder.** Miccosukee Gooseberry. Lake margins. Jefferson County. Federally and state endangered.

Hamamelidaceae (Witchhazel Family)

1. First pair of lateral veins coterminous with the leaf margin for more than 1 mm, flowers creamy or yellowish white, borne in terminal, oblong clusters, appearing in late spring..*Fothergilla gardenii*
1. First pair of lateral veins not coterminous with the leaf margin, or coterminous for less than 1 mm, flowers yellow, borne in axillary clusters, appearing fall, winter, or early spring
   2. Leaves glabrous or sparsely pubescent on the lower surface.....*Hamamelis virginiana*
   2. Leaves densely and conspicuously pubescent, the pubescence easily discerned by touch..*Hamamelis mollis*

!*Fothergilla gardenii* **Linnaeus.** Dwarf Witchalder. Flatwoods, shrub bogs, margins of drains in sandy pinelands. Central and western panhandle. State endangered.
*Hamamelis mollis **Oliver.** Chinese Witchhazel. Seepage slopes in mesic pinelands. Washington County. China.
Hamamelis virginiana **Linnaeus.** American Witchhazel. Mesic slopes, wet or moist stream margins, moist or wet hammocks. Panhandle and northern peninsula, south to Highlands County.

Hydrangeaceae (Hydrangea Family)

1. Plant a high-climbing woody vine, typically tightly appressed to a supporting tree, sometimes scrambling on the ground..*Decumaria barbara*
1. Plant a shrub, never vinelike
   2. Inflorescence usually with both sterile and fertile flowers (occasionally with all fertile flowers), the sterile flowers much the largest, usually long-stalked (in comparison to fertile flowers), borne well outward from the fertile flowers, and with 4 showy, petaloid sepals
      3. Most leaves conspicuously lobed..*Hydrangea quercifolia*
      3. All leaves unlobed, but coarsely toothed............................*Hydrangea arborescens*
   2. All flowers fertile and of the same size

4. Stamens twice as many as the petals; leaf hairs stellate...................*Deutzia scabra*

4. Stamens more than twice as many as the petals; leaf hairs not stellate...................
....*Philadelphus inodorus*

***Decumaria barbara* Linnaeus.** Climbing Hydrangea, Woodvamp, Cowitch Vine. Moist or wet hammocks and slopes, swamps and swamp margins. Panhandle south to central peninsula.

****Deutzia scabra* Thunberg.** Fuzzy Pride-of-Rochester. Escaped from cultivation. Franklin County. Asia.

!***Hydrangea arborescens* Linnaeus.** Wild Hydrangea, Mountain Hydrangea. Rich, moist slopes and bluffs. Liberty and Walton counties. State endangered.

***Hydrangea quercifolia* W. Bartram.** Oakleaf Hydrangea, Seven-Bark, Graybeard. Bluffs, ravines, usually where limestone is present.

***Philadelphus inodorus* Linnaeus.** Mockorange, Summer Dogwood, English Dogwood, Syringa. Rich woods along drainages, calcareous hammocks.

Hypericaceae (St. John's-wort Family)

1. Leaves narrow, linear, needlelike, the margins essentially parallel, not exceeding about 1 mm broad
   2. Largest leaves on main stem usually less than 12 mm long
      3. Plants low, often decumbent, mat forming, branches often upright from a reclining stem, usually not exceeding about 50 cm tall, young stems angled in cross section, capsule 6–10 mm long............................*Hypericum reductum*
      3. Plants taller, erect, usually 50–150 cm tall, not mat forming, young twigs flattened in cross section, capsule 3.5–5.5 mm long................*Hypericum brachyphyllum*
   2. Largest leaves on main stem usually greater than 12 mm long
      4. Bark of larger stems and upper trunk smooth, with a metallic luster; lower portion of trunk stripping in curly plates; plants often with an overall grayish or glaucous cast; seeds at least 1.5 mm long.............................*Hypericum lissophloeus*
      4. Bark of larger stems and trunk roughened, exfoliating in irregular strips; plants green; seeds less than 1 mm long
         5. Plant usually single-stemmed or with few branches near the top, stem limber, flexible, at maturity usually less than 1 m tall and not exceeding about 8 mm diameter near ground level... *Hypericum exile*
         5. Plants usually diffusely branched, stems not limber, somewhat stiff, at maturity usually exceeding 1 m tall and 1 cm diameter near ground level
            6. Bark reddish, cinnamon, or buff, exfoliating in thin sheets
               7. Bark of lower stem spongy and soft, easily compressed when squeezed, often exceeding 3 cm thick, trunk diameter at base often greater than 7

cm...*Hypericum chapmanii*
      7. Bark of lower stem only slightly if at all spongy and soft, usually firm if
        squeezed, trunk diameter usually less than 7 cm...*Hypericum fasciculatum*
    6. Bark grayish, grayish brown, or blackish gray, tight, thin, if exfoliating,
      doing so in flakes or narrow strips.....................*Hypericum nitidum*
1. Leaves not needlelike, broader, many or most at least (1.5) 2 mm broad
    8. Leaves mostly less than 1 cm long, many or most leaves less than 3 mm broad
      9. Leaves 5–10 mm long, 1.5–2 mm broad, appearing 4-ranked when viewed down
        the branch from the branch tip, sepals narrow and about the same size; a
        rounded shrub to about 1m tall................................*Hypericum microsepalum*
      9. Leaves 4–8 mm long, 2–3 mm broad, leaves not appearing conspicuously
        4-ranked, outer 2 sepals broad, conspicuous, inner 2 sepals tiny, scarcely evident;
        plant very low-growing, usually less than 15 cm tall........*Hypericum suffruticosum*
    8. Leaves predominantly longer, broader, or both
      10. Mature leaves predominantly 3–6 cm long
        11. Leaves separating cleanly from the stem when pulled gently downward, not
          simultaneously removing a strip of subtending bark....*Hypericum frondosum*
        11. Leaves not separating cleanly from the stem when pulled gently downward,
          usually retaining a short, narrow strip of bark at the base of the petiole
          12. Flowers usually in many-flowered cymes terminating the branches, the
            flowers far exceeding 8 in number; capsule 4–5 mm long....................
            ....*Hypericum nudiflorum*
          12. Flowers usually in simple, 3-flowered cymes, or if in compound cymes,
            then flowers usually 8 or fewer; capsule 8–10 mm long......................
            ....*Hypericum apocynifolium*
      10. Most or all mature leaves not exceeding 3 cm long
        13. Leaves broadly ovate, heart-shaped, clasping or nearly so at base
          14. Sepals and petals 4....................................*Hypericum tetrapetalum*
          14. Sepals and petals 5....................................*Hypericum myrtifolium*
        13. Leaves of various shapes, but not as above
          15. Sepals and petals 4
            16. Styles 2, inner sepals tiny and inconspicuous, appearing absent.........
            ....*Hypericum hypericoides*
            16. Styles 3–4, inner sepals smaller than outer sepals, lance shaped,
              conspicuous
              17. Leaves gland-dotted below (requires magnification), outer sepals
                acuminate at the apex, bark grayish or nearly black, tight, smooth.
                ....*Hypericum edisonianum*
              17. Leaves gland-dotted on both surfaces (requires magnification),
                outer sepals rounded or blunt at the apex, bark reddish brown,
                exfoliating in thin strips or flakes..........*Hypericum crux-andreae*
        15. Sepals and petals 5
          18. Leaves separating cleanly from the stem when pulled gently
            downward, not simultaneously removing a narrow strip of subtending

bark

19. Capsule 6–19 mm long, the apex with an elongated beak............
....*Hypericum prolificum*
19. Capsule 3–8 mm long, the apex lacking an elongated beak
....*Hypericum galioides*
18. Leaves not separating cleanly from the stem when pulled gently downward, usually retaining a narrow strip of bark at the base of the petiole...….......................…...................*Hypericum cistifolium*

Hypericum apocynifolium **Small.** Early St. John's-wort. Moist ravine slopes, bluffs, ravine bottoms. Central panhandle, Walton to Gadsden and Liberty counties.

Hypericum brachyphyllum **(Spach) Steudel.** Coastalplain St. John's-wort. Wet ditches, margins of swamps and pineland depressions, moist or wet flatwoods, pond margins. Nearly throughout, except the Keys.

•*Hypericum chapmanii* **P. Adams.** Spongebark Hypericum, Apalachicola St. John's-wort. Swamp and bog margins, flatwoods depressions, margins of wet flatwoods. Panhandle, Liberty and Franklin counties west to Santa Rosa County.

Hypericum cistifolium **Lamarck.** Roundpod St.John's-wort. Wet flatwoods, savannas, margins of coastal marshes, moist roadsides, interdune swales, margins of bogs and swamps. Nearly throughout, except the Keys.

Hypericum crux-andreae **(Linnaeus) Crantz.** St. Peter's-wort. Moist and wet flatwoods, savannas, bog and swamp margins, roadsides, well-drained uplands. Panhandle, east to Nassau County, sporadically south to Collier County, mostly absent from the east coast south of St. Johns County.

•!*Hypericum edisonianum* **(Small) W. P. Adams & N. Robson.** Arcadian St. John's-wort, Edison's St. John's-wort, Edison Ascyrum. Central peninsula scrub, Polk to Collier counties. State endangered.

•*Hypericum exile* **W. P. Adams.** Florida Sands John's-wort. Central panhandle, Bay, Gulf, Franklin, Liberty, and Leon counties.

Hypericum fasciculatum **Lamarck.** Sandweed, Peelbark St. John's-wort. Flatwoods, bog, swamp and pond margins, ditches bordering cypress ponds and swamps. Nearly throughout, except the Keys.

Hypericum frondosum **Michaux.** Golden John's-wort, Cedarglade St. John's-wort. Elevated river banks, temperate hardwood forests. Jackson and Gadsden counties.

Hypericum galioides **Lamarck.** Bedstraw St. John's-wort. Floodplains, bottoms, wet stream margins, wet hammocks, savannas and flatwoods, swamp margins. Panhandle, east to Nassau County, sporadically south to Hillsborough County.

Hypericum hypericoides **(Linnaeus) Crantz.** St. Andrew's-cross. Upland woods, pinelands, wet and dry flatwoods, sandhills, pond and swamp margins. Nearly throughout, except the Keys.

•!*Hypericum lissophloeus* **W. P. Adams.** Smoothbark St. John's-wort, Smoothbark Hypericum. Margins of sandy ponds and sinks. Central panhandle, Bay and Washington counties. State endangered.

Hypericum microsepalum **(Torrey & A. Gray) A. Gray ex S. Watson.** Flatwoods St.

John's-wort. Moist and wet flatwoods. Panhandle, Walton to Madison, Lafayette, and Dixie counties.

Hypericum myrtifolium **Lamarck.** Myrtleleaf St. John's-wort. Pond, bog, and swamp margins, wet ditches and roadsides, flatwoods, wet depressions in pinelands. Nearly throughout, except the Keys.

Hypericum nitidum **Lamarck.** Carolina St. John's-wort. Wet flatwoods, margins of blackwater streams, pond margins. Panhandle, Escambia to Madison counties, mostly nearer the coast.

Hypericum nudiflorum **Michaux ex Willdenow.** Early St. John's-wort. River and stream banks, wet hammocks, bottomlands, floodplains. Central panhandle, Walton to Leon and Wakulla counties.

Hypericum prolificum **Linnaeus.** Shrubby St. John's-wort. Sandy hammocks. Volusia County.

Hypericum suffruticosum **W. P. Adams & N. Robson.** Pineland St. John's-wort. Moist flatwoods, well-drained pinelands, sandhills. Panhandle, Escambia to Liberty and Gulf counties; northeast peninsula, Suwannee to Nassau, south to Alachua and St. Johns counties.

Hypericum tenuifolium **Pursh.** Atlantic St. John's-wort. Moist margins of sandy ponds, scrub, sandhills. Panhandle, Escambia to Franklin counties; peninsula, Duval and Alachua to Palm Beach and Collier counties.

Hypericum tetrapetalum **Lamarck.** Fourpetal St. John's-wort. Flatwoods, pond margins, wet roadsides, and ditches. Okaloosa and Nassau counties, south to Miami-Dade County.

Illiciaceae (Anisetree Family)

1. Flowers conspicuous, petals numerous (> 20), red, 15–25 mm long...............................
....*Illicium floridanum*
2. Flowers inconspicuous, petals few, yellow or greenish yellow, 4–5 mm long.................
....*Illicium parviflorum*

Illicium floridanum **J. Ellis.** Florida Anise. Moister parts of seepage slopes, swamp margins, bottomlands along rivers. Panhandle, from about Wakulla County westward.

•!*Illicium parviflorum* **Michaux ex Ventenat.** Ocala Anise, Yellow Anise, Star Anise. Central peninsula, Marion County south to Polk County, east to Volusia County. State endangered. Widely planted for ornament within and beyond Florida.

Iteaceae (Sweetspire Family)

Itea virginica **Linnaeus.** Virginia Sweetspire, Virginia Willow. Swamps, wet flatwoods, stream margins, wet woodlands. Panhandle south to Collier and Broward counties.

Juglandaceae (Walnut or Hickory Family)

1. Leaves odd-pinnate, terminal leaflet about the same size as or larger than the lateral leaflets, husk of fruit hard, splitting at maturity, usually somewhat longer than broad, usually not exceeding 6 cm long, or if rounded, less than 4 cm in diameter
    2. Leaflets of most leaves numbering 9–17
        3. Leaflets commonly 9 or 11 (varying 7–17), fruit (with husk attached) 2.5–4 cm long, usually conspicuously flattened in cross-section, plants occurring naturally on natural levees and floodplains of larger rivers..............................*Carya aquatica*
        3. Leaflets commonly 11 or more in number (varying 5–17), fruit (with husk attached) 3.5–5 cm long, more or less ellipsoid or slightly flattened in cross-section, the pecan of commerce; plants occurring in disturbed upland habitats, where usually established from previous plantings.....................*Carya illinoiensis*
    2. Leaflets usually 9 or fewer
        4. Leaflets usually 7 or more in number (varying 3–9)
            5. Fruit 4–6 cm long, husk 8–10 mm thick...........................*Carya tomentosa*
            5. Fruit 1.5–4 cm long, husk to 4 mm thick
                6. Mature fruit more or less round, predominantly 1.5–2.5 cm in diameter........
                ....*Carya cordiformis*
                6. Mature fruit oblong or obovoid, predominantly longer than 2.5 cm.............
                ....*Carya pallida*
        4. Leaflets not exceeding 7 in number (varying 3–7)
            7. Buds densely covered with small, rusty, resinous, granular scales; plants of white sand scrub or xeric sandhills in the central peninsula.......*Carya floridana*
            7. Buds covered with short hairs, lacking rusty, resinous glands; plants typically of dry or moist hammocks and mixed woodlands......................*Carya glabra*
1. Leaves, or most of them, often appearing evenly pinnate, the terminal leaflet absent or much reduced, husk of the fruit not splitting at maturity, more or less rounded, usually 5–8 cm in diameter...*Juglans nigra*

Carya aquatica **(Michaux f.) Nuttall.** Water Hickory. Floodplains, elevated levees along large rivers. Panhandle and peninsula, Escambia County east to Duval County, south to Lee, Hendry, and Palm Beach counties.

Carya cordiformis **(Wangenheim) K. Koch.** Bitternut Hickory, Yellowbud Hickory. Upland and lowland woods, moist hammocks, floodplains. Central panhandle, Holmes, Washington, and Bay counties east to Leon County.

•*Carya floridana* **Sargent.** Florida Hickory, Scrub Hickory. Scrub, sandhills. Marion and Volusia counties south to Charlotte and Miami-Dade counties.

Carya glabra **(Miller) Sweet.** Pignut Hickory. Moist slopes, dry uplands, bottomlands, wet or moist hammocks, often where limestone is present. Panhandle and peninsula, Escambia to Nassau counties, south to Charlotte, Highlands, and Brevard counties.

*Carya illinoinensis **(Wangenheim) K. Koch.** Pecan. Fields, disturbed sites, roadsides. Sporadically naturalized from Escambia to Orange counties; absent from the east coast. South-central U.S., probably in the vicinity of the Mississippi River.

Carya pallida **(Ashe) Engler & Graebner.** Sand Hickory. Sandhills, dry uplands, dry slopes. Panhandle, Santa Rosa to Leon counties.

Carya tomentosa **(Poiret) Nuttall.** Mockernut Hickory. Dry upland woods and slopes, dry hammocks. Panhandle and northern peninsula, Escambia to Columbia counties, south and east to Citrus, Sumter, and Volusia counties.

Juglans nigra **Linnaeus.** Black Walnut. Rich woods and slopes, ravines, bottoms, floodplains, river banks. Panhandle, sporadically from Escambia to Leon and Wakulla counties.

Lamiaceae or Labiatae (Mint Family)

1. Leaves palmately compound with 3–9 leaflets
  2. Leaves mostly with 3 leaflets, lower surface densely whitish or grayish hairy, middle leaflet sessile or nearly so..*Vitex trifolia*
  2. Leaves with 3–9 leaflets, predominately 5 or 7, lower surface of leaf, if hairy, not grayish or whitish, middle leaflet with a distinct stalk
    3. Lower surface of leaflet glabrous except for distinct tufts of hairs in the axils of the midvein and lateral veins..*Vitex glabrata*
    3. Lower surface of leaflet hairy throughout
      4. Leaflets mostly with 3–5 leaflets.....................................*Vitex negundo*
      4. Leaflets mostly with 5–7 leaflets...............................*Vitex agnus-castus*
1. Leaves simple
  5. Leaves narrow, linear, linear-oblong, oblong, elliptic, or obovate, length not exceeding about 3.5 cm
    6. Functional anther bearing stamens 2...........................*Hedeoma graveolens*
    6. Functional anther bearing stamens 4
      7. Flowers borne in congested, terminal, headlike spikes...........*Piloblephis rigida*
      7. Flowers borne in axillary clusters

8. Anthers with hornlike projections
  9. Flower petals of a single, solid color, lacking spots, lines, or other patterning...*Dicerandra immaculata*
  9. Flowers conspicuously decorated with darker spots or broken lines
    10. Flowers purplish, pink, or rose
      11. Anthers yellow, style glabrous or nearly so..*Dicerandra cornutissima*
      11. Anthers lavender, style with stiff hairs..........*Dicerandra thinicola*
    10. Flowers white or cream
      12. Anthers lavender or white
        13. Inflorescence with 1–2 open flowers, corolla usually white...*Dicerandra frutescens* var. *frutescens*
        13. Inflorescence with 2–3 open flowers, corolla more often pinkish white, turning pink with age..*Dicerandra modesta*
      12. Anthers bright yellow...........................*Dicerandra christmanii*
8. Anthers lacking hornlike projections
  14. Lower surface of the leaves glabrous or with erect hairs; branches of the stigma distinctly unequal in length
    15. Flowers red, 2–2.5 cm long........................*Calamintha coccinea*
    15. Flowers white, lavender, or pink, 1–2 cm long
      16. Leaf blade strongly revolute......................*Calamintha ashei*
      16. Leaf blade flat
        17. Leaves distinctly petiolate, glabrous...*Calamintha georgiana*
        17. Leaves short-stalked or sessile, minutely and densely hairy...*Calamintha dentata*
  14. Lower surface of the leaves with appressed hairs, branches of the style equal or nearly so
    18. Calyx tube glabrous or finely and inconspicuously hairy...*Conradina glabra*
    18. Calyx tube densely and conspicuously hairy
      19. Leaves with 2–4 lateral veins.....................*Conradina etonia*
      19. Leaves lacking lateral veins
        20. Midrib on lower surface of leaf densely hairy, density of hairs about equal to that of the blade tissue; lower lip of corolla 4–9 mm long......................................*Conradina canescens*
        20. Midrib on lower surface of leaf glabrous or moderately hairy, density of hairs less than that of the blade tissue; lower lip of corolla 9–16 mm long.....................*Conradina grandiflora*
5. Leaves broader, mature leaves longer than 3.5 cm
  21. Flowers borne in terminal panicles
    22. Flowers to about 5 mm long, individually not showy.........*Premna odorata*
    22. Flowers 10 mm long or longer, usually showy
      23. Corolla tube greater than 5 cm long..................*Clerodendrum indicum*
      23. Corolla tube not exceeding about 3 cm long

24. Plant vinelike, if shrubby then trailing and climbing......................
....*Clerodendrum* × *speciosum*
24. Plant a shrub or small tree
   25. Calyx at least 1 cm long
      26. Inflorescence densely crowded, corolla often double............
....*Clerodendrum chinense*
      26. Inflorescence not densely crowded, corolla single.............
....*Clerodendrum trichotomum*
   25. Calyx less than 1 cm long
      27. Leaves densely hairy on both surfaces............................
....*Clerodendrum speciosissimum*
      27. Leaves sparsely hairy to nearly glabrous on one or both
      surfaces
         28. Largest leaves greater than 15 cm long, to at least 15 cm
         broad
            29. Inflorescence a cyme, 4–9 cm long......................
....*Clerodendrum bungei*
            29. Inflorescence a panicle, 15–35 cm long..................
....*Clerodendrum kaempferi*
         28. Largest leaves less than 15 cm long, to about 5 cm broad
....*Clerodendrum glabrum*
21. Flowers borne in axillary clusters
   30. Flowers with 2 functional anthers...........................*Cornutia grandifolia*
   30. Flowers with 4 functional anthers
      31. Flowers pink, tiny, less than about 4 mm long.........*Callicarpa americana*
      31. Flowers slightly or considerably larger, not pink
         32. Corolla 20–25 mm long, bright red............*Holmskioldia sanguinea*
         32. Corolla 4–6 mm long, greenish white................*Petitia domingensis*

!*Calamintha ashei* (**Weatherby) Shinners.** Ashe's Calamint, Ashe's Calamintha.
Sandhills, scrub. Central peninsula ridge, Marion and Volusia to Glades counties. State
threatened.
Calamintha coccinea (**Nuttall ex Hooker) Bentham.** Red Basil, Scarlet Calamint.
Scrub, sandhills, dunes, deep sands, essentially near the coast. Panhandle (Escambia to
Wakulla counties) and central peninsula (Citrus and Brevard south to Hillsborough and
Indian River counties).
!*Calamintha dentata* **Chapman.** Florida Calamint, Toothed Savory. Sandhills, bluff mar-
gins. Central panhandle, Walton to Gadsden and Wakulla counties. State threatened.
!*Calamintha georgiana* (**R. M. Harper) Shinners.** Georgia Calamint. Bluff margins,
roadsides. Escambia and Holmes counties. State endangered.
Callicarpa americana **Linnaeus.** American Beautyberry, French Mulberry. Sandhills,
pinelands, dry woodlands, disturbed sites, flatwoods, hammocks. Throughout, including
the Keys.
*Clerodendrum bungei **Steudel.** Rose Glorybower. Disturbed sites, escaped from cultiva-

tion. Sporadically and potentially naturalized statewide, Escambia to Duval and Miami-Dade counties. China.

Clerodendrum chinense (**Osbeck**) **Mabberley.** Stickbush. Disturbed sites, escaped from cultivation. Sporadically naturalized statewide, mostly southern peninsula from Martin and Collier counties southward, except the Keys. Asia.

Clerodendrum glabrum **E. Meyer.** Natal Glorybower. Disturbed sites, escaped from cultivation. Miami-Dade County. China.

Clerodendrum indicum (**Linnaeus**) **Kuntze.** Turk's Turban, Skyrocket. Disturbed sites, escaped from cultivation. Sporadically and potentially naturalized statewide, except the Keys. East Indies.

Clerodendrum kaempferi (**Jacquin**) **Siebold ex Steudel.** Japanese Glorybower, Kaempfer's Glorybower. Disturbed sites, escaped from cultivation. Naturalized in Volusia and Miami-Dade counties, potentially elsewhere. Asia.

Clerodendrum speciosissimum **Van Geert ex C. Morren.** Javanese Glorybower. Disturbed sites, escaped from cultivation. Southern peninsula, Brevard, Polk, and Lee counties southward, including the Keys.

Clerodendrum trichotomum **Thunberg.** Harlequin Glorybower. Disturbed sites, escaped from cultivation. Naturalized in Escambia and perhaps Jackson counties. China.

Clerodendrum × speciosum **Dombrain.** Java Glory Bean, Pagoda Flower. Disturbed sites, escaped from cultivation. Sporadically naturalized in the central peninsula, Alachua County south to Hillsborough and Palm Beach counties.

Conradina canescens **A. Gray.** False Rosemary, Minty Rosemary, Wild Rosemary. Beaches, dunes, sandhills. Panhandle (Escambia to Franklin counties) and central peninsula (Hernando, Polk, and Highlands counties). Plants of the central peninsula population have previously been treated as *C. brevifolia* Shinners.

•!*Conradina etonia* **Kral & McCartney.** Etonia Rosemary, Etonia False Rosemary. Scrub. Putnam County. Federally and state endangered.

•!*Conradina glabra* **Shinners.** Apalachicola False Rosemary. Sandhills. Santa Rosa and Liberty counties. Federally and state endangered.

•!*Conradina grandiflora* **Small.** Largeflower False Rosemary. Coastal and inland scrub. East coast, Volusia County to Miami-Dade County. State threatened.

Cornutia grandifolia (**Schlechtendal & Chamisso**) **Schauer.** Azulejo. Disturbed sites, pine rocklands, escaped from cultivation. Miami-Dade County. Tropical America.

•!*Dicerandra christmanii* **Huck & Judd.** Lake Wales Balm, Christman's Mint. Scrub. Highlands County. Federally and state endangered.

•!*Dicerandra cornutissima* **Huck.** Longspur Balm, Robin's Mint. Scrub. Marion and Sumter counties. Federally and state endangered.

•!*Dicerandra frutescens* **Shinners.** Scrub Balm, Lloyd's Mint. Scrub and sandhills. Highlands County. Federally and state endangered.

•!*Dicerandra immaculata* **Lakela.** Lakela's Mint, Olga's Mint. Scrub. Indian River and St. Lucie counties. Federally and state endangered.

•!*Dicerandra immaculata* **Lakela var.** *savannarum* **Huck.** Savanna Balm, Dicerandra-of-the-Savannas. Coastal dunes and interdune swales. St. Lucie County.

•!*Dicerandra modesta* (**Huck**) **Huck.** Blushing Scrub Balm. Scrub. Polk County. Federally and state endangered.

•!*Dicerandra thinicola* **H. A. Miller.** Titusville Balm. Scrub. Brevard County. State endangered.

•!*Hedeoma graveolens* **Chapman ex A. Gray.** Mock Pennyroyal. Flatwoods, sandhills, roadsides, pond margins. Panhandle, Bay to Leon and Wakulla counties. State endangered.

Holmskioldia sanguinea **Retzius.** Chinese Hat. Disturbed sites, escaped from cultivation. Miami-Dade County. Asia, East Indies.

Petitia domingensis* **Jacquin. Bastard Stopper. Disturbed sites, escaped from cultivation. Miami-Dade County.

Piloblephis rigida **(W. Bartram ex Bentham) Rafinesque.** Wild Pennyroyal. Pinelands, dry uplands, sandhills, flatwoods. Peninsula, Duval, Alachua, and Citrus counties southward, except Monroe County and the Keys.

*Premna odorata **Blanco.** Fragrant Premna. Disturbed sites, escaped from cultivation. Miami-Dade County. Taiwan, Philippines.

Vitex agnus-castus* **Linnaeus. Lilac Chastetree, Chastetree. Disturbed sites, escaped from cultivation. Sporadically and potentially naturalized nearly statewide, Escambia to Collier counties. Europe.

*Vitex glabrata **R. Brown.** Smooth Chastetree. Disturbed sites, escaped from cultivation. Miami-Dade County. Australia.

*Vitex negundo **Linnaeus.** Negundo Chastetree. Disturbed sites, escaped from cultivation. Hernando County. Asia.

*◆*Vitex trifolia* **Linnaeus.** Simpleleaf Chastetree. Disturbed sites, escaped from cultivation. Largely confined to coastal counties of the central and southern peninsula, Pinellas and Volusia counties southward, including the Keys. Asia. FLEPPC listed (II).

Lauraceae (Laurel Family)

1. Plants evergreen
    2. Leaves or many of them with two major lateral veins, these more prominent than the other lateral veins, arising just above the base of the blade, and often running at least 70% the length of the blade
        3. Leaves alternate, even if closely set.........................*Cinnamomum camphora*
        3. Leaves, or many of them, opposite..................................*Cinnamomum verum*
    2. Lateral veins of the leaves equally prominent, usually distinctly pinnate
        4. Fertile stamens 3; fruit subtended by a distinctive, double-rimmed cup..............
        ....*Licaria triandra*
        4. Fertile stamens 9; fruit, if subtended by a cup, the cup not double-rimmed
            5. Outer 3 tepals of the flower distinctly shorter than the inner 3; fruit subtended by a woody cupule...*Ocotea coriacea*

5. Tepals of equal length; fruit not subtended by a woody cup
  6. Fruit usually well over 5 cm long, the avocado of commerce....................
    ....*Persea americana*
  6. Fruit usually not exceeding about 2 cm long
    7. Lower surface and midrib of leaf with long, wavy, shaggy brownish hairs; stalk of the infructescence usually conspicuously elongated.................
      ....*Persea palustris*
    7. Lower surface of leaf lacking erect, brownish hairs; hairs appressed and often difficult to see without magnification; stalk of the infructescence shorter
      8. Lower surface of the leaf usually whitish, not silky to the touch; hairs difficult to see without magnification.......................*Persea borbonia*
      8. Lower surface of the leaf usually rusty or coppery, silky to the touch.....
      ....*Persea humilis*
1. Plants deciduous
  9. Leaf blades 1–3 cm long, to about 1 cm broad..........................*Litsea aestivalis*
  9. Most leaves well over 3 cm long and greater than 1 cm broad
    10. Blades of at least some leaves (sometimes few) variously 2–3-lobed, usually with at least a few leaves mittenlike in outline...................*Sassafras albidum*
    10. Leaf blade unlobed
      11. Leaf blade leathery, usually not exceeding about 8 cm long, faintly aromatic when crushed..*Lindera subcoriacea*
      11. Leaf blade papery thin, few to many exceeding 8 cm long, strongly aromatic when crushed
        12. Leaf blade obovate, tapered at base, upper surface often quilted in appearance, with moderately depressed veins; leaves at tip of twig usually appreciably larger than those below; plant a shrub or small tree exceeding 2 m tall..................................*Lindera benzoin*
        12. Leaf blade elliptical, oval, or oblong, rounded at base, upper surface not quilted in appearance; leaves at tip of twig not appreciably larger than those below; plant a shrub to about 2 m tall, often shorter.................
        ....*Lindera melissifolia*

*◆*Cinnamomum camphora* (Linnaeus) J. Presl.** Camphortree. Disturbed uplands, secondary woods. Panhandle and peninsula, Escambia to Duval counties, south to St. Lucie and Lee counties. Eastern Asia. FLEPPC listed (I)

Cinnamomum verum* J. Presl. Cinnamon. Disturbed sites. St. Lucie County. China.

!*Licaria triandra* (Swartz) Kostermans. Gulf Licaria, Pepperleaf Sweetwood. Subtropical hammocks. Miami-Dade County. State endangered.

***Lindera benzoin* (Linnaeus) Blume.** Spicebush, Northern Spicebush. Moist or wet hammocks, streambanks, alluvial woods, moist wooded slopes. Central panhandle (Jackson, Calhoun, Gadsden, Liberty counties), central east peninsula (Putnam, Orange, Brevard counties).

!*Lindera melissifolia* (Walter) Blume. Pondberry, Southern Spicebush. Pond margins, bogs, sandy sinks, depression swamps. Collected in Florida without notation of county; presumably Gadsden County. Federally and state endangered.

!*Lindera subcoriacea* **B. E. Wofford.** Bog Spicebush. Seepage and streamside bogs. Escambia and Okaloosa counties. State endangered.

!*Litsea aestivalis* **(Linnaeus) Fernald.** Pondspice. Margins of cypress ponds and wetlands, swamp margins. Panhandle and northern peninsula, Okaloosa to Baker and St. Johns counties, south to Marion and Pasco counties. State endangered.

Ocotea coriacea **(Swartz) Britton.** Lancewood. Hammocks and hammock margins. East-central, southeastern, and southern peninsula, Volusia to Monroe and Miami-Dade counties, including the Keys.

*Persea americana **Miller.** Avocado. Disturbed sites, hammocks. Sporadically naturalized in Hernando, Lee, and Miami-Dade counties, presumably elsewhere. Tropical America.

Persea borbonia **(Linnaeus) Sprengel.** Red Bay. Dry or moist upland woods, dunes, rich slopes, sandy hammocks. Essentially throughout, including the Keys.

•*Persea humilis* **Nash.** Silk Bay. Scrub. Central peninsula, Levy, Alachua, and Putnam counties south to Collier and Martin counties.

Persea palustris **(Rafinesque) Sargent.** Swamp Bay. Swamps, bayheads, wet flatwoods, savannas. Essentially throughout, including the Keys.

Sassafras albidum **(Nuttall) Nees von Esenbeck.** Sassafras. Upland mixed woods, hammocks, pinelands, fields, fencerows, roadsides. Panhandle and northern peninsula, Escambia to Baker and Clay counties, south to Hillsborough and Orange counties.

Leitneriaceae (Corkwood Family)

!*Leitneria floridana* **Chapman.** Corkwood. Wet depressions, margins of freshwater and brackish marshes, often in association with limestone. Big Bend, largely near the coast, Franklin to Levy counties. State threatened.

Lythraceae (Loosestrife Family)

1. Leaves opposite or whorled, flowers borne in axillary clusters, plant an aquatic shrub...
....*Decodon verticillatus*
1. Leaves alternate, flowers borne in terminal racemes, plant a multistemmed shrub or
   small tree..*Lagerstroemia indica*

***Decodon verticillatus* (Linnaeus) Elliott.** Swamp Loosestrife, Willow-herb. Freshwater marshes, swamps. Panhandle and northern peninsula, Escambia to Duval counties, south to Hardee, DeSoto, and Highlands counties.
****Lagerstroemia indica* Linnaeus.** Crapemyrtle. Disturbed sites. Sporadically naturalized, Escambia County, east to Suwannee and Marion counties, south to Manatee and Highlands counties. Asia, Northern Australia.

Magnoliaceae (Magnolia Family)

1. Leaves more or less squarish; truncate, rounded, or cordate at base; apex truncate,
   usually broadly and shallowly notched; fruit an aggregate of samaras.....................
   ....*Liriodendron tulipifera*
1. Leaves not as above; fruit an aggregate of follicles
   2. Plant evergreen
      3. Leaves thick, distinctly leathery, lustrous dark green above; pale green, brownish,
         or rusty beneath...*Magnolia grandiflora*
      3. Leaves thinner, moderately leathery, pale or medium green above, chalky whitish
         or strongly glaucous beneath.......................*Magnolia virginiana* var. *australis*
   2. Plant deciduous
      4. Most leaves eared at base (auriculate)
         5. Leaves chalky white beneath, larger blades usually much exceeding 20 cm long
            ....*Magnolia ashei*
         5. Leaves greenish beneath, or if glaucous, then not chalky white, larger blades
            usually not exceeding 20 cm long............................*Magnolia pyramidata*
      4. Leaves not eared at base
         6. Blades of larger leaves 20–60 cm long, elliptic or elliptic-obovate, tapering
            gradually to the base from just below mid-blade; bud purplish, to about 3 cm
            long...*Magnolia tripetala*
         6. Blades of larger leaves 10–25 cm long, oblong, ovate, or oblong-obovate,
            tapering more or less abruptly to the base from well below mid-blade; bud
            silvery hairy, to about 2 cm long..............................*Magnolia acuminata*

Liriodendron tulipifera **Linnaeus.** Tuliptree, Tulip Poplar, Yellow Poplar. Bottomlands, bogs, swamp margins, streamside wetlands. Panhandle (Escambia to Leon and Wakulla counties) and northern peninsula (Nassau south to Pasco, Lake, and Orange counties).

!*Magnolia acuminata* **(Linnaeus) Linnaeus.** Cucumber Magnolia, Cucumbertree, Cowcumber. Rich slopes with temperate hardwood forests. Central panhandle, Holmes and Walton counties. State endangered.

•!*Magnolia ashei* **Weatherby.** Ashe Magnolia. Ravine slopes, temperate hardwood forests. Panhandle, Santa Rosa east to Leon and Wakulla counties. State endangered.

Magnolia grandiflora **Linnaeus.** Southern Magnolia, Bullbay. Temperate hardwood hammocks, moist slopes and bottoms, dunes. Panhandle and northern peninsula, Escambia to Nassau counties, south to Sarasota and Glades counties.

!*Magnolia pyramidata* **W. Bartram.** Pyramid Magnolia. Ravine slopes, temperate hardwood hammocks. Panhandle, Santa Rosa County east to Leon County. State endangered.

!*Magnolia tripetala* **(Linnaeus) Linnaeus.** Umbrella Magnolia. Bluffs and ravine slopes. Panhandle, Santa Rosa, Okaloosa, and Gadsden counties. State endangered.

Magnolia virginiana **Linnaeus** var. *australis* **Sargent.** Sweetbay, Sweetbay Magnolia. Swamps, bayheads, wet hammocks, bogs, wet flatwoods, savannas, creek and ravine bottoms. Essentially throughout, except the Keys.

Malpighiaceae (Malpighia Family)

1. Plant a woody vine or climbing, vinelike shrub; fruit winged, resembling a samara
  2. Apex of leaf rounded or minutely notched; wings of the fruit to 2 cm long.............
  ....*Stigmaphyllon emarginatum*
  2. Apex of leaf acuminate; wings of the fruit at least 3 cm long....*Hiptage benghalensis*
1. Plant a shrub; fruit a capsule or drupe
  3. Fruit a capsule, sepals lacking glands...................................*Galphimia gracilis*
  3. Fruit a drupe, sepals with 2 basal glands
    4. Leaf margins entire, the majority of leaves averaging 3 cm long or less; filaments hairy at base (requires magnification), style tapering at the apex........................
    ....*Byrsonima lucida*
    4. Leaf margins spiny or entire, but if entire, the majority of leaves at least 5 cm long; filaments glabrous, style thickened at the apex
      5. Leaves 1–2 cm long, margins spiny, resembling a holly.....*Malpighia coccigera*
      5. Leaves at least 5 cm long, margins entire...................*Malpighia emarginata*

!*Byrsonima lucida* (**Miller**) **de Candolle.** Long Key Locustberry. Hammocks, pine rocklands. Miami-Dade county and the Keys. State threatened.

Galphimia gracilis* **Bartling. Slender Goldenshower, Thryallis, Golden Thryallis, Shower-of-Gold. Disturbed sites, escaped from cultivation. Miami-Dade County. Mexico, Guatemala.

Hiptage benghalensis* (Linnaeus**) **Kurz.** Hiptage. Disturbed sites, escaped from cultivation. Broward County. Southeast Asia, Philippines.

Malpighia coccigera* **Linnaeus. Miniature Holly. Disturbed sites, escaped from cultivation. Polk County. West Indies.

Malpighia emarginata* **Sessé y Lacasta & Moçiño ex de Candolle. Barbados Cherry. Disturbed sites, escaped from cultivation. Southern peninsula, potentially Lee to Miami-Dade counties. Tropical America.

Stigmaphyllon emarginatum* (Cavanilles**) **A. Jussieu.** Monarch Amazonvine. Disturbed sites, escaped from cultivation. Brevard County. West Indies.

Malvaceae (Hibiscus Family)

1. At least some leaves distinctly lobed
    2. Leaves 10–30 cm long, 3–5-lobed, petiole 20–50 cm long; petals absent, flowers expanding at fruiting into radiating follicles that open and become leaflike, with seeds exposed...*Firmiana simplex*
    2. Leaves usually less than 10 cm long, usually 3-lobed or fewer, petiole to about 10 cm long; petals conspicuous, showy
        3. Calyx subtended by a whorl of bracts often appearing as a second calyx; petals white, fading to yellow or pink, sometimes with a red spot at the base, 3–6 cm long...*Gossypium hirsutum*
        3. Calyx not subtended by a whorl of bracts; petals yellow, lacking a red spot at base, narrow, not exceeding about 1 cm long
            4. Capsule hairy...*Triumfetta rhomboidea*
            4. Capsule glabrous...*Triumfetta semitriloba*
1. Leaves predominately unlobed (sometimes undulate or toothed, rarely indistinctly or obscurely 3-lobed)
    5. Leaves predominately pinnately veined, with at most 4 veins arising from a single point at the base of the blade
        6. Leaves ovate; larger leaves more than 10 cm long (5–20 cm), greater than 5 cm broad
            7. Many leaves strongly inequilateral at base; apex short acuminate; flowering and fruiting clusters subtended by a conspicuous, elongated, winglike bract...........
            ....*Tilia americana*
            7. Leaves more or less equal and rounded or cordate at base; apex strongly

acuminate, with a narrow, elongated tip; flowering and fruiting clusters not as above...*Pavonia paludicola*

6. Leaves linear, lanceolate, narrowly oblong, oblong-elliptic, or narrowly ovate; larger leaves less than 10 cm long, usually not exceeding about 4 cm broad; flowering and fruiting clusters not as above

    8. Leaf margins entire

        9. Leaf blades linear or narrowly lanceolate, apex acute, to about 4 cm long...*Cienfuegosia yucatanensis*

        9. Leaf blade ovate, apex strongly acuminate with a narrow, elongated tip...*Pavonia paludicola*

    8. Leaf margins toothed

        10. Calyx subtended by a cluster of bracts, often appearing as a second calyx

            11. Flower red, remaining closed at anthesis, the staminal column extending well beyond the petals

                12. Lower surface of the leaf hairy...*Malvaviscus arboreus* var. *drummondii*

                12. Lower surface of leaf glabrous.............*Malvaviscus penduliflorus*

            11. Flower yellow, opening at anthesis, the staminal column about as long as the petals

                13. Fruit segments ornamented with 3 barbed awns to 1 cm long; leaf margins distinctly and coarsely toothed.................*Pavonia spinifex*

                13. Fruit segments lacking awns; leaf margins entire or remotely and obscurely toothed.................................*Pavonia paludicola*

        10. Calyx not subtended by a cluster of bracts

            14. Corolla pink, purple, or white; capsule with 4–10 seeds

                15. Inflorescence borne mostly opposite the leaf; fruit capsule distinctly sharply angled.................................*Melochia pyramidata*

                15. Inflorescence borne mostly in the leaf axil; angles of the fruit capsule rounded.................................*Melochia tomentosa*

            14. Corolla yellow or orange-yellow

                16. Stamens 3 or 5...*Waltheria indica*

                16. Stamens 10 or more

                    17. Stamens coalesced into a single staminal column...*Sida ulmifolia*

                17. Stamens free

                    18. Leaf densely hairy with stellate pubescence; fruit to 1.5 cm long...*Corchorus hirsutus*

                    18. Leaf hairless or nearly so; fruit 2.5–6.5 cm long...*Corchorus siliquosus*

5. Leaves predominately palmately veined, at least 5 veins arising from a single point at the base of the blade (often also with lateral veins arising from the midvein above)

    19. Calyx subtended by a cluster of bracts, often appearing as a second calyx

        20. Fruit of 5 radially arranged 1-seeded carpels that separate from each other at maturity

21. Fruit segments ornamented with 3 barbed awns to 1 cm long; leaf
    margins distinctly and coarsely toothed......................*Pavonia spinifex*

21. Fruit segments lacking awns; leaf margins entire or remotely and
    obscurely toothed.......................................*Pavonia paludicola*

20. Fruit a capsule or leathery and indehiscent

  22. Fruit leathery

    23. Corolla fully open at anthesis, yellow or creamy yellow with a purple
      spot at base when fresh, fading to reddish or purple.....................
      ....*Thespesia populnea*

    23. Corolla remaining closed at anthesis, red

      24. Lower surface of the leaf hairy.....................................
      ....*Malvaviscus arboreus* var. *drummondii*

      24. Lower surface of leaf glabrous...........*Malvaviscus penduliflorus*

  22. Fruit a dehiscent capsule, splitting along the midrib of each locule

    25. Stipular scar inconspicuous and not encircling the stem; leaf margin
      sharply or bluntly toothed

      26. Leaf surface densely hairy or rough to the touch

        27. Bracts below the calyx forked at apex......*Hibiscus furcellatus*

        27. Bracts below the calyx not forked at apex

          28. Corolla 2–3 cm long, red.................*Hibiscus poeppigii*

          28. Corolla 6–12 cm long, pink...........*Hibiscus grandiflorus*

      26. Leaf surface glabrous or nearly so

        29. Bracts subtending the calyx divided at the tip....................
        ....*Hibiscus acetosella*

        29. Bracts subtending the calyx not divided at the tip

          30. Petal margins entire, bracts subtending the calyx at least
          1/3 as long as the calyx......................................
          ....*Hibiscus rosa-sinensis* var. *rosa-sinensis*

          30. Petal margins finely cut, bracts subtending the calyx less
          than 1/10 the length of the calyx.............................
          ....*Hibiscus rosa-sinensis* var. *schizopetalus*

    25. Stipular scar conspicuous, encircling the stem; leaf margin entire or
      obscurely toothed

      31. Corolla wholly yellow...
      ....*Talipariti tiliaceum* var. *pernambucense*

      31. Corolla yellow with a red center...................................
      ....*Talipariti tiliaceum* var. *tiliaceum*

19. Calyx not subtended by a cuplike cluster of bracts

  32. Petiole often nearly as long as the blade, many on any plant 4 cm long or
    longer...*Abutilon permolle*

  32. Petiole to about 1.5 cm long, much shorter than the blade.....*Grewia asiatica*

Abutilon permolle (**Willdenow**) **Sweet.** Coastal Indian Mallow. Disturbed sites. Southern peninsula, mostly Collier and Broward counties southward, including the Keys; disjunct to Manatee County.

!*Cienfuegosia yucatanensis* **Millspaugh.** Yucatan Flymallow, Yellow Hibiscus. Salt marshes, coastal hammocks. Keys. State endangered.

*Corchorus hirsutus **Linnaeus.** Jackswitch, Woolly Corchorus, Cadillo. Disturbed sites, escaped from cultivation. Miami-Dade County. Tropical America.

Corchorus siliquosus **Linnaeus.** Slippery Burr. Disturbed sites. Southern peninsula, Collier and Miami-Dade counties southward, including the Keys.

*Firmiana simplex (**Linnaeus**) **W. Wight.** Chinese Parasol Tree, Varnish Tree. Disturbed sites, roadsides, mixed woodlands, escaped from cultivation. Panhandle, northern peninsula; Escambia to Leon counties; Putnam County. Subtropical China.

!*Gossypium hirsutum* **Linnaeus.** Wild Cotton, Upland Cotton. Tropical thickets, coastal hammocks, beaches, disturbed sites. Mostly southern peninsula, Pinellas and Palm Beach counties southward; escaped from cultivation in Gilchrist County. State endangered.

*Grewia asiatica **Linnaeus.** Phalsa. Disturbed pinelands. Miami-Dade County. India, Nepal, Cambodia, Laos, Thailand.

*Hibiscus acetosella **Welwitsch ex Hieronymus.** African Rosemallow. Disturbed sites, escaped from cultivation. Coastal counties of the central and southern peninsula, Volusia, Pinellas, and Hillsborough counties south to Miami-Dade County, absent from the Keys.

Hibiscus furcellatus **Desrousseaux.** Lindenleaf Rosemallow. Dry, open sites. East- and south-central peninsula, Brevard County south to Broward County.

Hibiscus grandiflorus **Michaux.** Swamp Rosemallow. Marshes, swamp margins, open wetlands. Nearly throughout, Escambia to Duval counties, south to Monroe and Miami-Dade counties; absent from the Keys.

!*Hibiscus poeppigii* (**Sprengel**) **Garcke.** Poeppig's Rosemallow, Wild Hibiscus. Hammocks, often in association with limestone. Miami-Dade County and the Keys. State endangered.

*Hibiscus rosa-sinensis **Linnaeus.** Garden Rosemallow, Shoe-Black-Plant. Disturbed sites, escaped from cultivation. Southeast coast, Brevard to Miami-Dade counties. Asia.

*Hibiscus rosa-sinensis **Linnaeus var.** *schizopetalus* **Dyer.** Fringed Rosemallow. Disturbed sites, escaped from cultivation. Southern peninsula; Lee, Collier, and Palm Beach counties. Africa.

*Malvaviscus arboreus **Cavanilles var.** *drummondii* (**Torrey & A. Gray**) **Schery.** Texas Waxmallow. Disturbed sites, escaped from cultivation. Sporadically naturalized, Leon and Franklin counties south to Miami-Dade County and the Keys. Mexico.

*Malvaviscus penduliflorus **de Candolle.** Mazapan, Turkscap Mallow. Disturbed sites, escaped from cultivation. Sporadically naturalized from Levy, Marion, and Volusia counties southward, including the Keys; also naturalized in Franklin County. Tropical America.

Melochia pyramidata **Linnaeus.** Pyramidflower. Disturbed sites. Southern peninsula, Broward and Monroe counties southward, including the Keys.

Melochia tomentosa **Linnaeus.** Woolly Pyramidflower, Teabush, Broomwood. Pine rocklands, Southeast peninsula, St. Lucie to Miami-Dade counties.

!*Pavonia paludicola* **Nicolson ex Fryxell.** Swampbush. Hammocks. Southern peninsula.

Collier, Monroe, Miami-Dade counties; absent from the Keys. State endangered.
Pavonia spinifex (**Linnaeus**) **Cavanilles.** Gingerbush. Hammocks. Northeast and central peninsula, Duval and Citrus counties south to Brevard and Highlands counties.
Sida ulmifolia **Miller.** Common Wireweed, Common Fanpetals. Disturbed sites, roadsides, margins of upland woods. Nearly throughout, including the Keys.
*◆*Talipariti tiliaceum* (**Linnaeus**) **Fryxell.** Sea Hibiscus, Mahoe. Disturbed sites, margins of coastal hammocks and woodlands. Southern peninsula, Manatee and St. Lucie counties southward, including the Keys. Tropical Asia. FLEPPC listed (II).
Talipariti tiliaceum* (Linnaeus**) **Fryxell var.** *pernambucense* (**Arruda**) **Fryxell.** Yellow Mahoe. Disturbed sites. Southern peninsula, sporadic from Lee and Miami-Dade counties southward, including the Keys. Tropical America.
*◆*Thespesia populnea* (**Linnaeus**) **Solander ex Correa.** Portia Tree, Seaside Mahoe. Disturbed sites, hammocks, beaches, roadsides. Central and southern peninsula, often in the coastal zone, Brevard and Sarasota counties southward, including the Keys. Old and New World tropics. FLEPPC listed (I).
Tilia americana (**Miller**) **Castiglioni.** American Bassword, Linden. Ravine slopes, mixed upland woods, temperate hardwood forests, moist hammocks. Panhandle and northern peninsula, Okaloosa to Flagler counties, south to Hillsborough, Polk, and Osceola counties. Includes *Tilia americana* Linnaeus var. *caroliniana* (Miller) Castiglioni and *Tilia americana* Linnaeus var. *heterophylla* (Ventenat) Loudon.
Triumfetta rhomboidea* **Jacquin. Diamond Burrbark. Disturbed sites, DeSoto and Broward counties. Tropical America.
Triumfetta semitriloba* **Jacquin. Sacramento Burrbark. Disturbed sites. Central and southern peninsula, Manatee and Okeechobee counties south to Miami-Dade County, absent from the Keys.
Waltheria indica **Linnaeus.** Waltheria, Buffcoat, Sleepy Morning. Open pinelands, sandhills, hammocks, disturbed sites. Central and southern Florida, Lake County southward, including the Keys.

Melastomataceae (Melastoma Family)

1. Upper surface of leaf glabrous, smooth to the touch; flowers white...*Tetrazygia bicolor*
1. Upper surface of leaf with appressed hairs, rough to the touch; flowers purple...........
....*Melastoma malabathricum*

Melastoma malabathricum* **Linnaeus. Straits Rhododendron, Malabar Melastome. Wet flatwoods. Martin County. Southeast Asia.
!*Tetrazygia bicolor* (**Miller**) **Cogniaux.** Tetrazygia, Florida Clover Ash. Pinelands, hammocks. Miami-Dade County. State threatened.

Meliaceae (Mahogany Family)

1. Leaves bi- or tripinnately compound, margins of leaflets toothed.........*Melia azedarch*
1. Leaves pinnately compound, margins of leaflets entire (sometimes wavy but not toothed)
    2. Blades of leaflets usually unequal at base and appearing recurved, with more tissue on one side of the midvein than the other...........................*Swietenia mahagoni*
    2. Blades of leaflets equal at base, not appearing recurved.............*Khaya senegalensis*

Khaya senegalensis **(Desrousseaux) A. Jussieu.** African Mahogany. Disturbed sites, hammocks, escaped from cultivation. Collier and Broward counties, presumably elsewhere. Tropical Africa.
*◆*Melia azedarach* **Linnaeus.** Chinaberry, Pride of India. Disturbed sites, upland woods. Essentially throughout, including the Keys. Asia. FLEPPC listed (II).
!*Swietenia mahagoni* **(Linnaeus) Jacquin.** Mahogany, West Indian Mahogany. Hammocks, disturbed sites. Southern peninsula, Lee and Broward counties southward, including the Keys. State threatened.

Menispermaceae (Moonseed Family)

1. At least some leaves peltate (petiole attachment often near the base of the blade).........
....*Menispermum canadense*
1. Leaves not peltate
    2. Bracts of the inflorescence minute; fruits glabrous; stamens 6 or 12; plants of central Florida northward
        3. At least some leaves moderately or deeply 3–5 lobed; sepals 6, petallike, true petals absent; stamens at least 12; fruit black at maturity, averaging about 20 mm long...*Calycocarpum lyonii*
        3. Leaves, at most, shallowly lobed or undulate; sepals and petals 6; stamens 6; fruit red at maturity, 6–8 mm long...*Cocculus carolinus*
    2. Bracts of the inflorescence large, conspicuous; fruits hairy; stamens 4; plants of subtropical hammocks in the southern peninsula, very rare (if extant)....................
....*Cissampelos pareira*

Calycocarpum lyonii **(Pursh) A. Gray.** Cupseed. Floodplains, wet hammocks, margins of upland woods, disturbed sites. Central panhandle, Washington and Jackson to Jefferson counties.

!*Cissampelos pareira* **Linnaeus.** Periera Brava. Subtropical hammocks. Miami-Dade County. State endangered.

Cocculus carolinus **(Linnaeus) de Candolle.** Carolina Moonseed, Redberry Moonseed, Carolina Coralbead, Snailseed, Coralbeads. Mixed woodlands, woodland margins, fencerows, moist hammocks. Panhandle and northern peninsula, Escambia to Duval counties, south to Citrus and Orange counties.

Menispermum canadense **Linnaeus.** Moonseed, Common Moonseed. Low woods, floodplains, moist hammocks. Central panhandle (Liberty to Jefferson counties), Alachua County.

Moraceae (Mulberry Family)

1. Leaf margins toothed, deeply lobed, or wavy
    2. Flowers not evident, produced inside of a fruitlike receptacle
        3. Leaf margins wavy, apex narrowly tapered and elongated into a taillike tip (caudate)..*Ficus religiosa*
        3. Leaf margins deeply lobed and coarsely, bluntly, or obscurely toothed..............
        ....*Ficus carica*
    2. Flowers evident (though often small), not produced inside of a fruitlike receptacle
        4. Pistillate flowers borne in a globose inflorescence; styles unbranched; leaves alternate, opposite, or whorled, all on the same plant, especially on young branches and new growth; young petioles and twigs with long hairs......................
        ....*Broussonetia papyrifera*
        4. Pistillate flowers borne in a cylindrical inflorescence; styles with 2 branches; leaves all alternate; young petioles and twigs glabrous or with short hairs
            5. Upper surface of the leaf blade glabrous or only sparsely hairy, more or less lustrous, usually smooth to the touch....................................*Morus alba*
            5. Upper surface of the leaf blade with short, appressed, stiff hairs pointing toward the tip of the blade, rendering the surface sandpapery to the touch (hairs sometimes sloughing with age, leaving only the pustular base and retaining the sandpapery feel); surface more or less dull green..................*Morus rubra*
1. Leaf margins entire
    6. Flowers not evident, completely enclosed within a fruitlike receptacle
        7. Plant a scrambling or trailing vine, attaching to the substrate by adventitious roots.
        ....*Ficus pumila*
        7. Plant an erect tree or shrub, even if germinating as an epiphyte
            8. Leaf margin often wavy; leaf apex elongated into a tail-like tip about 1/2 the length of the blade...*Ficus religiosa*
            8. Leaf margin entire; leaf apex obtuse or short-pointed, the point much shorter than 1/2 the length of the blade
                9. Leaf blade distinctly obovate or reverse deltoid, the apex rounded..............
                ....*Ficus deltoidea*

9. Leaf blade elliptic or ovate, the apex pointed
  10. Many leaves with 3–4 pairs of basal veins (some with only 2 pairs); fruit hairy...…...........*Ficus benghalensis*
  10. Basal pairs of leaf veins never more than 2; fruit glabrous
    11. Leaf blade with more than 10 uniform lateral veins, these usually uniformly and regularly spaced
      12. Leaf blade averaging 4–6 cm long (potentially to 11 cm); stipule less that 1.5 cm long; fruit nearly globose...........*Ficus benjamina*
      12. Leaf blade 9–30 cm long; stipule at least 3 cm long; fruit oblong or ovoid..…......*Ficus elastica*
    11. Leaf blade with fewer than 10 lateral veins, or if more than 10, then the veins not uniformly and regularly spaced
      13. Fruit sessile or essentially so, the fruit stalk never more than 5 mm long
        14. Blade of fully developed leaves 6 cm long or longer; fruit usually at least 1 cm in diameter, 1.7–2.8 cm long
          15. Basal pair of leaf veins often V-shaped, their angle from the midvein often distinctly different than the angle of the veins above..............................…...............*Ficus altissima*
          15. Angle from the midvein of the basal pair of veins about equal to the angle of those above.................…......*Ficus aurea*
        14. Blade of many leaves less than 6 cm long; fruit 5–6 mm in diameter, 9–11 mm long..........…................*Ficus microcarpa*
      13. Fruit stalked
        16. Petiole averaging 1.5–12 cm long (a few as short as 7 mm).........…...….....................*Ficus citrifolia*
        16. Petiole 1 cm long or less...............................…......*Ficus americana*
6. At least some flowers evident, not completely enclosed within a fruitlike receptacle
  17. Leaves ovate or lanceolate, not leathery, plants deciduous; "fruit" round, knobby, 8–12 cm in diameter, appearing orangelike (technically an aggregate of red or orange achenes enclosed within expanded calyces)...............…......*Maclura pomifera*
  17. Leaves oblong, leathery, plants evergreen; "fruit" about 1.5 cm in diameter........…...….................*Brosimum alicastrum*

Brosimum alicastrum **Swartz.** Breadnut. Disturbed sites, escaped from cultivation. Miami-Dade County and the Keys. Tropical America.

◆Broussonetia papyrifera **(Linnaeus) Ventenat.** Paper Mulberry. Disturbed sites, often near human habitations and commercial esblishments, escaped from cultivation. Asia. FLEPPC listed (II).

◆Ficus altissima **Blume.** False Banyan, Council Tree, Lofty Fig. Hammocks, disturbed sites. Southern peninsula, Collier and Palm Beach counties southward, except the Keys. China, India, Malaysia, southeast Asia. FLEPPC listed (II).

Ficus americana **Aublet.** West Indian Laurel Fig. Hammocks, disturbed sites. Miami-Dade County. West Indies, tropical America.

Ficus aurea **Nuttall.** Strangler Fig. Hammocks, margins of mangrove swamps. Central and southern peninsula, Volusia and Pinellas counties southward, including the Keys.

Ficus benghalensis **Linnaeus.** Banyan Tree. Hammocks, disturbed sites. Broward and Miami-Dade counties, perhaps elsewhere. India, Pakistan.

Ficus benjamina **Linnaeus.** Weeping Fig. Disturbed sites. Sporadically naturalized in the southern peninsula, St. Lucie and Lee counties southward, including the Keys.

Ficus carica **Linnaeus.** Common Fig. Disturbed sites, escaped from cultivation. Sporadically and sparingly naturalized, Gulf to Miami-Dade counties. Asia Minor.

Ficus citrifolia **Miller.** Wild Banyan Tree. Hammocks. Essentially the southernmost peninsula, Collier and Broward counties southward, including the Keys; disjunct to Hillsborough County.

Ficus deltoidea **Jack.** Mistletoe Rubberplant. Disturbed sites. Brevard County. Malaysia.

Ficus elastica **Roxburgh ex Hornemann.** India Rubberplant. Disturbed sites. Broward County. India.

◆Ficus microcarpa **Linnaeus f.** Laurel Fig, Indian Laurel. Disturbed sites, lake margins, floodplain swamps, rockland hammocks. Southern peninsula, Hillsborough, Highlands, and Martin counties southward, including the Keys. Asia. FLEPPC listed (I).

Ficus pumila **Linnaeus.** Climbing Fig. Disturbed sites, building facades. Sporadically and sparingly naturalized, Escambia to Lake and Pinellas counties. Southern Asia.

Ficus religiosa **Linnaeus.** Bo Tree, Sacred Fig. Disturbed sites, hammocks. Miami-Dade County and the Keys. India, Southeast Asia.

Maclura pomifera **(Rafinesque) C. K. Schneider.** Osage Orange, Hedge Apple. Disturbed sites. Sporadically and sparingly naturalized. Escambia to Marion and Volusia counties. South-central U.S.

Morus alba **Linnaeus.** White Mulberry. Disturbed sites, upland woods. Sporadically but regularly naturalized, Escambia to Duval counties, south to Brevard and Lee counties. East Asia.

Morus rubra **Linnaeus.** Red Mulberry. Upland woods, coastal and inland hammocks, floodplains. Essentially throughout, except the Keys.

Moringaceae (Horseraddishtree Family)

Moringa oleifera **Lamarck.** Horseradishtree. Disturbed sites, escaped from cultivation. Sparsely naturalized, Manatee and Miami-Dade counties and the Keys, perhaps elsewhere. India, Sri Lanka.

Muntingiaceae (Muntingia Family)

*__Muntingia calabura__ Linnaeus.** Strawberrytree. Tropical hammocks and pinelands, usually where disturbed. Southern peninsula, Collier, Hendry, and Palm Beach counties southward, including the Keys.

Myoporaceae (Myoporum Family)

*__Bontia daphnoides__ Linnaeus.** White Alling. Shorelines, salt flats, margins of mangrove swamps, coastal thickets. Miami-Dade County. West Indies, northern South America.

Myricaceae (Wax Myrtle Family)

1. Leaves entire, not aromatic when crushed...................................*Myrica inodora*
1. Leaves toothed, at least toward the apex, aromatic when crushed
  2. Lower and upper leaf surfaces finely punctate with amber or brown glands
    3. Mature plants well over 1 m tall, mature leaves usually greater than 4 cm long, blades mostly oblanceolate; habit a large shrub or small tree.................................
    ....*Myrica cerifera* var. *cerifera*
    3. Mature plants 1 m tall or less, mature leaves usually less than 4 cm long, blades narrower, at least some more nearly linear lanceolate; habit a dwarf shrub...........
    ....*Myrica cerifera* var. *pumila*
  2. Lower leaf surface finely punctate with amber or brown glands (requires magnification to see clearly)...*Myrica caroliniensis*

Myrica caroliniensis **Miller.** Swamp Candleberry, Northern Bayberry, Evergreen Bayberry. Swamps, wet flatwoods and flatwoods depressions, bogs, wet pine savannas. Panhandle (Escambia to Leon and Wakulla counties), northeast peninsula (Nassau to Marion counties), central peninsula (Polk and Highlands counties), perhaps elsewhere.
Myrica cerifera **Linnaeus var. *cerifera*.** Wax Myrtle, Bayberry, Southern Bayberry. Upland woods, pinelands, ravine slopes, mixed hardwood forests, hammocks, moist flatwoods, margins of coastal marshes. Essentially throughout, including the Keys.
Myrica cerifera **Linnaeus var. *pumila* Michaux.** Dwarf Wax Myrtle. Flatwoods, pinelands. Presumably throughout the state.
Myrica inodora **W. Bartram.** Odorless Bayberry. Bogs, swamps, wet flatwoods, wetland margins. Panhandle, Escambia to Leon and Wakulla counties.

Myrsinaceae (Myrsine Family)

1. Inflorescence fasciculate, the stalk of the inflorescence shorter than the petiole...........
....*Myrsine cubana*
1. Inflorescence not fasciculate, the stalk of the inflorescence longer than the petiole
  2. Plant a stoloniferous sub-shrub, usually not exceeding about 30 cm tall; leaf margin minutely or coarsely toothed; inflorescence 3–5-flowered.............*Ardisia japonica*
  2. Plant a shrub or small tree, usually much exceeding 1 m tall; leaf margin conspicuously toothed or entire; inflorescence 5–18-flowered (or more)
    3. Leaf margin toothed...*Ardisia crenata*
    3. Leaf margin entire
      4. Inflorescence strictly axillary or axillary and terminal
        5. Petiole 10–20 mm long...*Ardisia solanacea*
        5. Petiole 5–10 mm long...*Ardisia elliptica*
      4. Inflorescence terminal.............................*Ardisia escallonioides*

*◆***Ardisia crenata*** **Sims.** Coral Ardisia, Scratchthroat. Hammocks, ravine slopes and bottoms, disturbed woodlands. Central panhandle to south-central peninsula, naturalizing readily and potentially elsewhere. Japan, southern Asia. FLEPPC listed (I).
*◆***Ardisia elliptica*** **Thunberg.** Shoebutton Ardisia. Hammocks. Southern peninsula. Brevard and Pinellas counties southward, especially the southernmost peninsula; absent from the Keys. Asia. FLEPPC listed (I).
Ardisia escallonioides **Schiede & Deppe ex Schlechtendal & Chamisso.** Marlberry. Hammocks, pinelands. Flagler and Pasco counties southward, including the Keys.
*◆***Ardisia japonica*** **(Thunberg) Blume.** Japanese Ardisia. Disturbed hammocks. Jackson and Alachua counties, presumably elsewhere. FLEPPC listed (II).
****Ardisia solanacea*** **Roxburgh.** China-shrub. Disturbed sites, escaped from cultivation. Hillsborough County.
Myrsine cubana **A. de Candolle.** Myrsine, Colicwood. Largely confined to the southern peninsula, Dixie and Volusia counties southward; Wakulla County.

Myrtaceae (Myrtle Family)

1. Leaves alternate, fruit a woody capsule
  2. Leaf venation pinnate; petals at first fused into a lidlike covering that separates at anthesis to expose numerous stamens
    3. Young leaves and twigs reddish hairy; apex of the leaf blunt.........................

....*Eucalyptus torelliana*

3. Young leaves and twigs glabrous; apex of the leaf long acuminate or long
tapering

    4. Bark essentially smooth throughout, exfoliating

        5. Fruit stalk flattened in cross section..........................…....……*Eucalyptus grandis*

        5. Fruit stalk rounded in cross section...

        ....*Eucalyptus camaldulensis* subsp. *acuta*

    4. Bark rough, exfoliating only apically...........................…*Eucalyptus robusta*

2. Leaf venation parallel, with 1–8 veins; petals not fused as described above

    5. Flowers white, leaves usually with 3 or more veins

        6. Leaves less than 1 cm broad....................................…...…*Melaleuca linariifolia*

        6. Leaves well over 1 cm broad (usually about 2.5 cm)...*Melaleuca quinquenervia*

    5. Flowers red, leaves usually with a single midvein...............*Callistemon viminalis*

1. Leaves opposite, fruit fleshy

    7. Lower surface of leaf whitish-tomentose; flowers rosy-pink............................

    ....…*Rhodomyrtus tomentosa*

    7. Lower surface of leaf glabrous, or if hairy, not whitish-tomentose; flowers, if
present, not rosy-pink

        8. Inflorescence a several- or many-flowered compound panicle

            9. Petals absent; calyx forming a lidlike structure (calyptra) that separates along
one side or falls off at anthesis to expose the stamens

                10. Inflorescence finely hairy...............................…....…*Calyptranthes pallens*

                10. Inflorescence glabrous...............................…...……*Calyptranthes zuzygium*

            9. Petals present; calyx not forming a lid

                11. Leaves broadly elliptic; flower buds 5–6 mm long, calyx lacking lobes,
petals united, forming a lid.........................…....…................…*Syzygium cumini*

                11. Leaves narrowly elliptic or lanceolate; flower buds 15–30 cm long, calyx
lobed, petals free...…....…..............*Syzygium jambos*

        8. Inflorescence a raceme or 3-flowered cluster, or flowers solitary

            12. Calyx closed or nearly so in bud, at anthesis splitting into 4 or 5 irregular or
unequal lobes

                13. Leaves 1–5 cm long, usually less than 2 cm long; flowers and fruit on a
stalk 2–3.5 cm long; fruit about 1 cm long..................…....…*Mosiera longipes*

                13. Leaves 4–14 cm long; flower and fruit stalks to about 2 cm long; fruit 2–4
cm long

                    14. Lower surface of leaf pubescent; veins conspicuously impressed above
and raised below; twigs angled...........................…....……*Psidium guajava*

                    14. Lower surface of leaf glabrous; veins not conspicuously impressed or
raised; twigs rounded in cross section.................*Psidium cattleianum*

            12. Calyx open in bud, with 4 nearly equal lobes

                15. Inflorescence a cyme..…....…*Myrcianthes fragrans*

                15. Inflorescence a raceme, sometimes congested and appearing fasciculate;
or flowers solitary

                    16. Leaves usually 8 cm long or longer; flowers 3–4 cm in diameter.........

....*Syzygium jambos*

16. Leaves usually not exceeding about 6 cm long, flowers not exceeding about 2 cm in diameter
    17. Flower stalk not exceeding 5 mm long
        18. Blade of most leaves oblanceolate, widest above the middle, the apex rounded...................................*Eugenia foetida*
        18. Blade of most leaves ovate or lanceolate, widest near the middle, the apex tapering to a distinct point....*Eugenia axillaris*
    17. Flower stalk more than 5 mm long
        19. Sepals usually 4 mm long or longer, lanceolate or long-elliptic, about equal in size and shape, often reflexed at maturity.........
        ....*Eugenia uniflora*
        19. Sepals usually not exceeding about 3.5 mm long, orbicular or ovate, in 2 unequal pairs
            20. Upper surface of leaf blade dull, apex broadly acuminate....
            ....*Eugenia rhombea*
            20. Upper surface of leaf blade lustrous, apex narrowly acuminate.................................*Eugenia confusa*

◆Callistemon viminalis (**Solander ex Gaertner**) **G. Don ex Loudon.** Bottlebrush, Weeping Bottlebrush. Disturbed sites, escaped from cultivation. Widely planted, sporadically naturalized, Highlands and Martin counties southward. Australia. FLEPPC listed (II).

!*Calyptranthes pallens* **Grisebach.** Pale Lidflower, Spicewood. Hammocks. Miami-Dade County and the Keys. State threatened.

!*Calyptranthes zuzygium* (**Linnaeus**) **Swartz.** Myrtle-of-the-River. Hammocks. Miami-Dade County and the Keys. State endangered.

Eucalyptus camaldulensis **Dehnhardt subsp.** *acuta* **Brooker & M. W. McDonald.** River Redgum. Disturbed sites. Charlotte County. Australia.

Eucalyptus grandis **W. Hill ex Maiden.** Grand Ecalyptus. Disturbed sites. Southeastern peninsula, Glades, Hendry, and Palm Beach counties; Pinellas County. Australia, Queensland to New South Wales.

Eucalyptus robusta **Smith.** Swamp Mahogany. Disturbed sites, sporadically naturalized. Central peninsula, mostly in coastal counties, Brevard County south to Martin County; Pinellas south to Lee County. Australia.

Eucalyptus torelliana **F. Mueller.** Torrell's Eucalyptus, Cadaga. Disturbed sites. Lee and Palm Beach counties, perhaps elsewhere. Northeast Australia.

Eugenia axillaris (**Swartz**) **Willdenow.** White Stopper. Coastal hammocks, inland hammocks in the southern peninsula. Central and southern peninsula, Levy and Volusia counties southward, including the Keys.

!*Eugenia confusa* **de Candolle.** Red-berry Stopper. Hammocks. Martin and Miami-Dade counties and the Keys. State endangered.

Eugenia foetida **Persoon.** Spanish Stopper, Boxleaf Stopper, Boxleaf Eugenia. Hammocks, pinelands. Central and southern peninsula, Manatee and Brevard counties, southward to the Keys.

!*Eugenia rhombea* **Krug & Urban ex Urban.** Red Stopper. Hammocks. Miami-Dade County and the Keys. State endangered.

*◆*Eugenia uniflora* **Linnaeus.** Surinam Cherry. Disturbed sites, hammocks. South-central and southern peninsula, Pinellas, Hillsborough, Polk, and Brevard counties southward, including the Keys. Brazil. FLEPPC listed (I).

*_Melaleuca linariifolia_ **Smith.** Cajeput Tree, Snow in Summer. Disturbed sites, wetlands. Osceola County. Eastern Australia.

*◆*Melaleuca quinquenervia* **(Cavanilles) S. T. Blake.** Cajeput, Punk Tree. Hammocks, pinelands, woodlands, disturbed sites. Central and southern peninsula, Hernando, Orange, and Brevard counties southward, including the Keys. Australia, Melaneisa. FLEPPC listed (I).

!*Mosiera longipes* **(O. Berg) Small.** Long-stalked Stopper, Mangroveberry. Hammocks, pinelands. Miami-Dade County and the Keys. State threatened.

!*Myrcianthes fragrans* **(Swartz) McVaugh.** Simpson's Stopper, Twinberry. Mostly coastal hammocks, occasionally inland hammocks in the southern peninsula. St. Johns County southward along the coast to Martin Coast; Lee and Broward counties, southward to the Keys. State threatened.

*◆*Psidium cattleianum* **Sabine.** Strawberry Guava. Disturbed sites, prairies, hydric hammocks, marl pariries, bottomland forests, bays, seepage slopes. Central and southern peninsula, Pinellas, Hillsborough, and Seminole counties southward to Miami-Dade County; absent from the Keys. Southeastern Brazil. FLEPPC listed (I).

*◆*Psidium guajava* **Linnaeus.** Guava. Disturbed sites, hammocks. Central and southern peninsula, Pinellas, Hillsborough, Polk, Osceola, and Brevard counties southward, including the Keys. Tropical America. FLEPPC listed (I).

*◆*Rhodomyrtus tomentosa* **(Aiton) Hasskarl.** Rose Myrtle. Flatwoods, pinelands, disturbed sites. South-central peninsula, Pasco, Orange, and Brevard counties south to Collier and Palm Beach counties. Asia, Australia. FLEPPC listed (I).

*◆*Syzygium cumini* **(Linnaeus) Skeels.** Java Plum. Disturbed sites, coastal hammocks, mesic flatwoods, rockland hammocks, lake margins. Central and southern peninsula, Hillsborough and Polk counties south to Collier and Broward counties. India, Malaysia, Southeast Asia. FLEPPC listed (I).

*◆*Syzygium jambos* **(Linnaeus) Alston.** Malabar Plum, Rose Apple. Disturbed sites, hammocks. Central and southern peninsula, Brevard and Sarasota counties southward, except the Keys. India, Malaysia, southeast Asia. FLEPPC listed (II)

Nyctaginaceae (Four-O'Clock Family)

1. Fruit more or less dry, angled, with 5 vertical rows of stalked glands, these sometimes only near the apex
    2. Plant a vine or sprawling vinelike shrub, stem armed with hooked spines..............
    ....*Pisonia aculeata*
    2. Plant a shrub or tree, stem not armed...............................…......*Pisonia rotundata*
1. Fruit fleshy, smooth, lacking glands
    3. Leaves thin, papery; petiole slender; buds and inflorescence sparsely hairy............
    ....….....*Guapira discolor*
    3. Leaves thick, leathery; petiole stout; buds and inflorescence densely hairy............
    ....…......*Guapira obtusata*

Guapira discolor (Sprengel) Little. Blolly, Beeftree. Hammocks, coastal scrub, pinelands. Southeast coast, Brevard, Monroe, and Miami-Dade counties and the Keys.
Guapira obtusata (Jacquin) Little. Broadleaf Blolly. Subtropical hammocks. Keys.
Pisonia aculeata Linnaeus. Devil's Claws, Pullback. Subtropical hammocks. Southern peninsula, Hillsborough, St. Lucie, and Glades counties southward, including the Keys.
!*Pisonia rotundata* Grisebach. Cockspur, Pisonia. Hammocks, scrub. Lower Keys. State endangered.

Nyssaceae (Tupelo Family)

1. Petioles of most leaves 3 cm long or longer....................................…......*Nyssa aquatica*
1. Petioles of most leaves not exceeding 3 cm long, often not exceeding about 2.5 cm long
    2. Largest leaves 10–30 cm long; fruit reddish, 2–4 cm long.................*Nyssa ogeche*
    2. Largest leaves usually not exceeding about 10 cm long; fruit purplish or blackish, not exceeding about 1.5 cm long
        3. Plant typically of upland habitats; many leaf blades papery to the touch, obovate, widest above the middle, often with 1 or 2 large marginal teeth near the apex; trunk usually not buttressed or conspicuously thickened at base......*Nyssa sylvatica*
        3. Plant typically of wetland habitats; most leaf blades somewhat thickened, elliptical or lance-elliptic; trunk often buttressed or conspicuously thickened at base
            4. Plant a tree, drupe more or less ellipsoid, often distinctly longer than broad; fruit stalk usually longer than 2 cm; most leaf blades 8 cm long or longer..........
            ....*Nyssa biflora*
            4. Plant a shrub, drupe more or less rounded, often shorter than broad; fruit stalk

to about 2 cm long; most leaf blades not exceeding about 7 cm long...............
....*Nyssa ursina*

***Nyssa aquatica* Linnaeus.** Water Tupelo. Floodplains, bayheads, river swamps, lake margins. Panhandle, Escambia to Madison and Levy counties.
***Nyssa biflora* Walter.** Blackgum. Swamps, pond margins, floodplains, bottomlands, bayheads. Panhandle and northern peninsula, south to about Lee, Glades, and St. Lucie counties.
***Nyssa ogeche* W. Bartram ex Marshall.** Ogeechee Tupelo, Ogeechee-lime. Floodplains, wet river margins, bayheads, river swamps, bottomlands. Panhandle and northern peninsula, Walton to Hamilton counties, south to Dixie, Alachua, and St. John's counties; cultivated and established in Hillsborough County.
***Nyssa sylvatica* Marshall.** Sour Gum, Upland Black Tupelo. Upland woods of various mixtures. Panhandle and northern peninsula, Escambia to Nassau counties, south to Sumter and Lake counties; Manatee County.
***Nyssa ursina* Small.** Bear Tupelo. Moist or wet flatwoods, swamps, sloughs. Central panhandle, Bay, Gulf, Liberty, Calhoun, Franklin, and Wakulla counties.

Oleaceae (Olive Family)

1. Leaves simple (or unifoliolate and appearing simple)
    2. Margin of leaf blade at least partially toothed, even if bluntly so
        3. Leaf blade acuminate at apex, the tip narrowly tapering to an elongated point......
        ....*Forestiera acuminata*
        3. Leaf blade blunt or rounded at apex, the tip at most broadly or bluntly tapered
            4. Pubescence between the leaf nodes on the current year's twigs confined to two lines, these on opposite sides of the twig; petiole glabrous or sparsely hairy; plant flowering in summer..*Forestiera ligustrina*
            4. Pubescence of the current year's twigs evenly distributed around the twig; petiole at least moderately pubescent; plant flowering in late winter or very early spring...*Forestiera godfreyi*
    2. Margin of leaf blade entire
        5. Few if any leaves exceeding 6 cm long
            6. Leaves sessile, the blade conspicuously oblanceolate or spatulate, tapering evenly and gradually toward the base.........................*Forestiera segregata*
            6. Leaves petiolate, the blade elliptic or oval, usually widest at or below the middle, or if some leaves wider above the middle, then not long tapering toward the base
                7. Tube of the corolla 1.5–3 times longer than the corolla lobes....................
                ....*Ligustrum ovalifolium*

7. Tube of the corolla about equaling in length the corolla lobes

    8. Individual flowers essentially stalkless....................*Ligustrum quihoui*

    8. Invididual flowers stalked.....................…..…………......*Ligustrum sinense*

5. Many or most leaves longer than 5 cm

  9. Twigs glabrous

    10. Plant evergreen; leaves thick, leathery; flower petals not linear, united about half their length

      11. Plant an erect, arching shrub or scrambling vinelike shrub or woody climber; leaves mostly 3–8 cm long

        12. Plant a scrambling, vinelike shrub or woody climber; flowers clustered in the leaf axils....................….......*Jasminum dichotomum*

        12. Plant a large ascending or arching shrub; flowers borne in terminal, many flowered panicles............................…....*Ligustrum japonicum*

      11. Plant a large shrub or small tree; many or most leaves well over 8 cm long

        13. Inflorescence axillary, few flowered; leaf blade flat or slightly revolute; corolla creamy- or greenish white

          14. Fruit oval or ellipsoid, not exceeding 1 cm in diameter; stone acute at both ends……....…....…...........*Osmanthus americanus*

          14. Fruit more or less rounded, 2–2.5 cm in diameter; stone acute only at the base.............................…....…....*Osmanthus megacarpus*

        13. Inflorescence terminal, many flowered; blade of many leaves folded upward from the midvein; corolla white or off-white.................…

        …………………………………………………………...…....*Ligustrum lucidum*

    10. Plant deciduous, leaves thinner, not leathery; flower petals linear, united only at the base

      15. Petals about 1 cm long; drupe 2–2.5 cm long…..*Chionanthus pygmaeus*

      15. Petals 2–3 cm long; drupe 1–1.5 cm long………*Chionanthus virginicus*

  9. Twigs pubescent

    16. Calyx lobes 1–2 mm long.............................…..........*Jasminum laurifolium*

    16. Calyx lobes 6–12 mm long

      17. Flowers double; apex of corolla lobes blunt.............*Jasminum sambac*

      17. Flowers not double; apex of corolla lobes pointed.........................…

      …………………………………………………....…….....*Jasminum multiflorum*

1. Leaves compound

  18. Plant a shrub or woody vine; fruit a berry

    19. Leaves with 5–7 leaflets.............................…................*Jasminum grandiflorum*

    19. Leaves with 3 leaflets

      20. Twigs glabrous; flowers yellow..............................…....*Jasminum mesnyi*

      20. Twigs pubescent; flowers white.........................…....*Jasminum fluminense*

  18. Plant a tree; fruit a samara

    21. Plants of well-drained uplands; leaves typically whitish beneath

      22. Samara 19–38 mm long, 3–6 mm broad; seed cavity not exceeding 11 mm long and 2.5 mm broad......................…...............*Fraxinus americana*

22. Samara 32–54 mm long, 5–8 mm broad; seed cavity of some or many
samaras exceeding 11 mm long (varying 7–15 mm long), 2–3.5 mm broad...
....*Fraxinus smallii*

21. Plants of wetland habitats, often where swampy; leaves typically greenish
beneath

23. Plant a medium or large tree; samaras narrow, 4–12 mm broad

24. Wing of the samara arising at about the middle of the seed cavity, the
cavity clearly evident; pubescence along the veins on the lower surface
of the leaflets absent, or short, narrowly banded, and not noticeably
tangled...*Fraxinus pennsylvanica*

24. Wing of the samara arising near the base of the seed cavity, the cavity
obscure; pubescence along the veins on the lower surface of the leaflets
longish, conspicuously banded, and noticeably tangled.....................
....*Fraxinus profunda*

23. Plant a large shrub or multi-stemmed tree; samaras often broader (except in
F. cubensis), many 12–30 mm broad

25. Samara rhombic, broadly elliptic, or elliptic obovate, usually 12–30 mm
broad; seed cavity usually at least 1/2 as long as the samara...............
....*Fraxinus caroliniana*

25. Samara elliptic-oblanceolate to narrowly elliptic-obovate, usually not
exceeding about 12 mm broad; seed cavity not exceeding 1/2 the length
of the samara

26. Lower surface of the leaflets coated with microscope rounded
protuberances that obscure the blade surface, minute peltate scales
not evident (all seen only with at least 20× magnification)............
....*Fraxinus pauciflora*

26. Lower surface of leaflets not coated as described above, minute
glandular peltate scales usually evident and abundant when viewed
with at least 20× magnification........................*Fraxinus cubensis*

•!*Chionanthus pygmaeus* **Small.** Pigmy Fringetree. Scrub. Central peninsula, Lake and
Seminole to Sarasota and Highlands counties. Federally and state endangered.
Chionanthus virginicus **Linnaeus.** Fringetree, Old-man's Beard, Grandsie-gray-beard.
Upland woods, mixed forests, moist hammocks, margins of floodplains, ravine slopes.
Panhandle, northern and central peninsula, Okaloosa to Duval counties, south to Sarasota,
Hardee, and Brevard counties.
Forestiera acuminata **(Michaux) Poiret.** Swamp Privet, Eastern Swampprivet. Flood-
plains and river swamps. Central panhandle in counties along the Apalachicola River,
northern peninsula chiefly along the Suwanne River; also reported from Hernando
County.
!*Forestiera godfreyi* **L. C. Anderson.** Godfrey's Privet, Godfrey's Swampprivet. Moist,
calcareous woods and hammocks. Central pahandle (Gadsden to Jefferson counties),
north-central peninsula (Gilchrist and Alachua to Sumter counties), Duval County. State
endangered.

Forestiera ligustrina (**Michaux**) **Poiret.** Upland Swampprivet. Upland woods, usually in association with limestone. Panhandle and northern peninsula, Escambia to Putnam and Pasco counties.

Forestiera segregata (**Jacquin**) **Krug & Urban.** Florida Privet, Florida Swampprivet. Coastal hammocks, scrub, thickets. Duval and Dixie counties southward, mostly along the coast, including the Keys.

Fraxinus americana **Linnaeus.** White Ash, American Ash. Rich upland woods, well-drained floodplains where water stands only briefly, hammocks. Panhandle and northern peninsula, Walton to Jefferson counties, Columbia to Hernando counties.

Fraxinus caroliniana **Miller.** Carolina Ash, Pop Ash, Water Ash. Swamps, floodplains, wet river banks, pond margins, usually where water stands much of the time. Escambia County east to Levy, Marion, and Flagler counties.

Fraxinus cubensis **Grisebach.** Hardwood hammocks, swamp forests, low woods, margins of ponds and lakes, river and creek banks. Marion to Monroe and Palm Beach counties; absent from the Keys.

Fraxinus pauciflora **Nuttall.** Sloughs, swamps, usually in standing water. Panhandle and northern peninsula, Walton to Nassau counties, south to Osceola County.

Fraxinus pennsylvanica **Marshall.** Green Ash, Red Ash. Swamps, floodplains. Panhandle and northern peninsula, south to about Brevard and Hillsborough counties.

Fraxinus profunda (**Bush**) **Bush.** Pumpkin Ash. River swamps, floodplains. Panhandle and northern peninsula, Santa Rosa to Marion, Osceloa, and Brevard counties.

Fraxinus smallii **Britton.** Bottomland forests, alluvial woods, creek terraces, floodplains, sandy swales, slopes, upland hardwoods. Reported for Florida in Leon County, presumably elsewhere.

*◆*Jasminum dichotomum* **Vahl.** Gold Coast Jasmine. Disturbed sites, scrub, mesic flatwoods, subtropical hammocks, escaped from cultivation. Southern peninsula, Highlands and St. Lucie counties, south to the Keys. Tropical Africa. FLEPPC listed (I)

*◆*Jasminum fluminense* **Vellozo.** Brazilian Jasmine, Azores Jasmin, Jazmin de Trapo. Disturbed sites, subtropical hammocks, escaped from cultivation. Southeastern peninsula, Highlands and St. Lucie counties south to Miami-Dade County and the Keys. Africa. FLEPPC listed (I).

Jasminum grandiflorum* **Linnaeus. Poet's Jasmine. Disturbed sites, escaped from cultivation. Palm Beach County. Eurasia.

Jasminum laurifolium* **Roxburgh ex Hornemann. Angelwing Jasmine. Disturbed sites, escaped from cultivation. Miami-Dade County. Papua New Guinea.

Jasminum mesnyi* **Hance. Japanese Jasmine. Disturbed sites, escaped from cultivation. Widely planted, sporadically naturalized, potentially statewide. China.

Jasminum multiflorum* (Burman f.**) **Andrews.** Star Jasmine. Disturbed sites, escaped from cultivation. Sproradically naturalized from Duval to Lee and Martin counties. India.

*◆*Jasminum sambac* (**Linnaeus**) **Aiton.** Arabian Jasmine. Disturbed sites, escaped from cultivation. Southern peninsula, Collier and Broward counties southward, including the Keys. India. FLEPPC listed (II).

Ligustrum japonicum* **Thunberg. Japanese Privet. Disturbed sites, mixed hardwood forests. Widely planted in the panhandle and northern peninsula, sporadically naturalized southward to Volusia County, perhaps elsewhere. Japan, eastern Asia.

*◆*Ligustrum lucidum* **Aiton f.** Glossy Privet, Wax-leaf Ligustrum, Tree Privet, Wax-leaf Privet. Disturbed sites, roadsides, mixed upland woods. Panhandle east to Duval County, south to Highlands County, presumably elsewhere. Asia. FLEPPC listed (I).

Ligustrum ovalifolium* **Hassk. California Privet. Disturbed sites, hammocks, escaped from cultivation. Sporadically naturalized, Duval and Hernando counties, presumably elsewhere. Japan.

Ligustrum quihoui* **Carrière. Waxleaf Privet. Disturbed sites, escaped from cultivation. Leon County. China.

*◆*Ligustrum sinense* **Loureiro.** Chinese Privet, Variegated Ligustrum. Hammocks, upland woods, bottomlands, floodplains, moist hammocks, temperate forests, stream banks, disturbed sites, escaped from cultivation. More common in the panhandle, Escambia to Suwannee counties, northern peninsula, south to Hillsborough County, also reported from Miami-Dade County. China. FLEPPC listed (I).

Osmanthus americanus **(Linnaeus) Bentham & Hooker f. ex A. Gray.** Wild Olive, Devilwood, American Devilwood. Inland and coastal hammocks, flatwoods, swamp margins, scrub, floodplains, wooded bluffs. Panhandle and northern peninsula, south to Manatee, Highlands, and Indian River counties.

•*Osmanthus megacarpus* **(Small) Small ex Little.** Scrub Wild Olive. Scrub. Central peninsula, Citrus, Marion, and Volusia counties south to Sarasota, DeSoto, Highlands, and Indian River counties.

Onagraceae (Evening Primrose Family)

1. Herbaceous portions of stem distinctly shaggy pubescent
    2. Sepals usually 4 (rarely 5), lower portion of the stem definitely woody................
    ....*Ludwigia peruviana*
    2. Sepals 5, stem mostly herbaceous, the lowest portion sometimes hard and
        woodlike...*Ludwigia leptocarpa*
1. Herbaceous portions of the stem glabrous or merely finely short-hairy....................
    ....*Ludwigia octovalvis*

Ludwigia leptocarpa **(Nuttall) H. Hara.** Anglestem Primrosewillow. Pond and swamp margins, ditches, calcareous hammocks. Essentially throughout the state, except the Keys.

Ludwigia octovalvis **(Jacquin) Raven.** Mexican Primrosewillow. Marshes, roadside ditches, dunal swales. Essentially throughout the state, including the Keys.

*◆*Ludwigia peruviana* **(Linnaeus) H. Hara.** Peruvian Primrosewillow. Marshes, roadside ditches, lake and pond margins. Essentially throughout the state, except the Keys. FLEPPC listed (I).

Passifloraceae (Passionflower Family)

1. Flowers subtended by a cluster of conspicuous, leaflike bracts..........*Passiflora edulis*
1. Flowers not subtended by a cluster of conspicuous, leaflike bracts
   2. Leaves densely, softly, and finely hairy; petals narrowly linear, 2–3 mm
     long..*Passiflora multiflora*
   2. Leaves glabrous or scarcely hairy; petals absent....................*Passiflora suberosa*

*Passiflora edulis **Sims.** Passionfruit, Purple Granadilla. Disturbed sites, escaped from cultivation. Southeastern peninsula, Palm Beach, Broward, and Miami-Dade counties, potentially elsewhere. South America (southern Brazil).

!Passiflora multiflora **Linnaeus.** Whiteflower Passionflower, White-flowered Passionvine. Subtropical hammocks. Miami-Dade County and the Keys. State endangered.

Passiflora suberosa **Linnaeus.** Corkystem Passionflower. Hammocks, shell middens, pinelands. Peninsula, Duval and Dixie counties south along the coast, generally throughout from Lee and Palm Beach counties southward, including the Keys.

Paulowniaceae (Princesstree Family)

Paulownia tomentosa **(Thunberg) Siebold & Zuccarini ex Steudel.** Princesstree. Disturbed sites, escaped from cultivation. Rarely naturalized, Gadsden and Alachua counties. Asia.

Pentaphylacaceae (Pentaphylax Family)

*Ternstroemia gymnanthera **(Wight & Arnott) Beddome.** Japanese Ternstroemia, Clayera. Dry, sandy secondary woods. Leon County. Asia. [Note: This species was discovered as this book was going to press. Time did not permit it or its family to be inserted into the keys.]

Phyllanthaceae (Phyllanthus Family)

1. Leaves compound with 3 leaflets..*Bichofia javanica*
1. Leaves simple
  2. Leaves multi-colored...…..….*Breynia disticha*
  2. Leaves wholly green or yellow green, sometimes whitish-green beneath
    3. Flowers with 5 tiny, rounded, greenish-white petals about 1 mm long; fruit
      capsule with 2 seeds per locule......................…..…............*Heterosavia bahamensis*
    3. Flower petals absent; fruit capsule with 1 seed per locule............................
      ....…............*Flueggea virosa* subsp. *melanthesoides*

*◆Bischofia javanica Blume. Bischofia, Bishopwood, Javanese Bishopwood. Hammocks, flatwoods, swamps, praries, disturbed sites, escaped from cultivation. Southern peninsula, Pinellas, Sarasota, and Martin counties southward; absent from the Keys. Asia, Malaysia. FLEPPC listed (I).
*Breynia disticha J. R. Forster & G. Forster. Snowbush. Disturbed sites, escaped from cultivation. Southern peninsula, potentially Lee and St. Lucie counties southward; presumably absent from the Keys.
*Flueggea virosa (Roxburgh ex Willdenow) F. Voigt subsp. *melanthesoides* (F. Mueller) G. L. Webster. Simpleleaf Bushweed. Disturbed sites, escaped from cultivation. Miami-Dade County. Old World tropics.
!Heterosavia bahamensis (Britton) Petra Hoffmann. Bahama Maidenbush. Coastal thickets, hammocks. Keys. State endangered.

Phytolaccaceae (Pokeweed Family)

1. Stamens 4; style present; berries red, orange, or yellow.....................…......*Rivina humilis*
1. Stamens 8–13; style absent; berries black.........................…....*Trichostigma octandrum*

Rivina humilis Linnaeus. Rougeplant, Bloodberry. Hammocks, disturbed sites. Peninsula, Duval, Alachua, and Levy counties southward, including the Keys.
!Trichostigma octandrum (Linnaeus) H. Walter. Hoopvine. Subtropical hammocks, shell middens. Southern peninsula, Collier and Broward counties southward, including the Keys. State endangered.

Picramniaceae (Bitterbush Family)

1. Leaves, or most of them, with more than 9 leaflets; leaflets 1–2.5 cm long, less than 2 cm broad..*Alvaradoa amorphiodes*
1. Leaves with 5–9 leaflets; leaflets 3–12 cm long, 2–5 cm broad......*Picramnia pentandra*

!*Alvaradoa amorphoides* **Liebmann.** Mexican Alvaradoa, Alvaradoa. Hammocks. Miami-Dade County. State endangered.
!*Picramnia pentandra* **Swartz.** Florida Bitterbush, Bitterbush. Hammocks. Miami-Dade County. State endangered.

Pinaceae (Pine Family)

1. Most or many leaves well over 11 cm long
  2. Sheath of most leaf fascicles at least 1.5 cm long; terminal bud silvery..................
    ....*Pinus palustris*
  2. Sheath of most leaf fascicles not exceeding 1.5 cm long; terminal bud brown
    3. Needles predominately 2 per fascicle (sometimes 2 and 3 per fascicle, at least some trees with leaves in fascicles of 2 and 3 well intermixed).........*Pinus elliottii*
    3. Needles predominately 3 per fascicle
      4. Sheath of most leaf fascicles about 1 cm long, mature cones rounded or top-shaped, 5–8 cm long, about as broad or broader than long, trunk often producing short adventitious branches for much of its extent......*Pinus serotina*
      4. Sheath of most leaf fascicles 1–1.5 cm long, mature cones elongate, usually longer than broad, 10–13 cm long.....................................*Pinus taeda*
1. Most or many leaves 11 cm long or less
  5. Leaves 2 per fascicle
    6. Leafless twigs smooth; bark lacking resin pockets
      7. Leaf sheath about 5 mm long, ovulate cone scales tipped inside with a conspicuous darkened band, plants usually of deep sand, more or less xeric habitats (often planted by forest product companies); bark pinelike..............
      ....*Pinus clausa*
      7. Leaf sheath 5–7 mm long, ovulate cone scales tipped inside with an inconcpicuous, pale brownish or tan band, or band lacking; plants of mesic woods, moist mixed hardwood hammocks, rich slopes, or the borders of bottomlands and floodplains; bark resembling a hardwood.........*Pinus glabra*
    6. Leafless twigs scaly; bark with conspicuous resin pockets...........*Pinus echinata*
  5. Leaves predominately 3 per fascicle (some leaves with 2 or 4 per fascicle)
    7. Sheath of most leaf fascicles about 1 cm long, mature cones rounded or top-

shaped, 5–8 cm long, about as broad or broader than long, trunk often producing short aventitious branches for much of its extent...................................*Pinus serotina*
7. Sheath of most leaf fascicles 1–1.5 cm long, mature cones elongate, usually longer than broad, 10–13 cm long...*Pinus taeda*

***Pinus clausa* (Chapman ex Engelmann) Vasey ex Sargent.** Sand Pine. Scrub, stable dunes, sandy woods. Panhandle (Escambia to Gadsden and Wakulla counties), peninsula (Suwannee south to Collier and Broward counties); widely planted by forest product companies beyond its normal habitat.
***Pinus echinata* Miller.** Shortleaf Pine, Shortneedle Pine, Yellow Pine. Upland woods, sandhills, often in association with clay lenses in the soil. Panhandle, Escambia to Columbia counties.
***Pinus elliottii* Engelmann.** Slash Pine. Flatwoods and margins of adjacent swamps. Throughout the state, including the Keys.
***Pinus glabra* Walter.** Spruce Pine. Rich moist woods, ravine bottoms, margins of floodplains. Panhandle and northern peninsula, Escambia to Baker, Union, Alachua, and Putnam counties.
***Pinus palustris* Miller.** Longleaf Pine. Sandhills, flatwoods. Panhandle and peninsula south to Lee, Glades, Okeechobee, and Indian River counties.
***Pinus serotina* Michaux.** Pond Pine. Wet flatwoods, swamp margins. Panhandle and northern peninsula, Okaloosa to Nassau counties, south to Polk, Osceola, and Brevard counties.
***Pinus taeda* Linnaeus.** Loblolly Pine. Low woodlands, mixed hardwood and pine forests, lower slopes of ravines, spreading into fields and disturbed sites. Panhandle and northern peninsula continuously south to Hernando, Sumter, Lake, and Orange counties; also reported from Hillsborough and DeSoto counties.

Piperaceae (Pepper Family)

1. Inflorescence a solitary spike borne opposite a leaf
  2. Base of leaf rounded or wedge-shaped; most or all leaves less than 11 cm broad......
  ....*Piper aduncum*
  2. Base of leaf distinctly cordate; most or all leaves greater than 11 cm broad..............
  ....*Piper auritum*
1. Inflorescence a cluster of several spikes borne in the leaf axil......*Lepianthes umbellata*

Lepianthes umbellata* (Linnaeus) Rafinesque. Baquina. Subtropical hammocks. Miami-Dade County. Tropical America.

*__Piper aduncum__ **Linnaeus.** Spiked Pepper, False Matico, Higuillo de Hoja Menuda. Subtropical hammocks, especially where disturbed. Miami-Dade and Monroe counties, except the Keys. Tropical America.

__Piper auritum__ **Kunth.** Veracruz Pepper. Subtropical hammocks, especially where disturbed. Broward and Miami-Dade counties. Tropical America.

Pittosporaceae (Pittosporum Family)

1. Leaf blade ovate or oblong, margins revolute; capsule rounded, to 1.2 cm diameter......
....*Pittosporum tobira*
1. Leaf blade obovate, the margins not revolute; capsule ellipsoid, not exceeding about
   1 cm diameter...*Pittosporum pentandrum*

*◆__Pittosporum pentandrum__ **(Blanco) Merrill.** Taiwanese Cheesewood. Disturbed sites, hammock margins. Miami-Dade County. Taiwan, Philippines. FLEPPC listed (II).

*__Pittosporum tobira__ **(Thunberg) Aiton.** Japanese Cheesewood. Hammocks. Widely planted, naturalized in Gadsden County. China, Japan.

Plantaginaceae (Plantain Family)

*__Russelia equisetiformis__ **Schlechtendal & Chamisso.** Fountainbush, Firecracker Plant. Disturbed sites, margins of hammocks and pinelands, escaped from cultivation. Southern peninsula, Pinellas and Brevard counties southward along the coast, generally throughout the southern peninsula, Lee, Hendry, and Palm Beach counties southward, including the Keys. Mexico.

Platanaceae (Planetree or Sycamore Family)

__Platanus occidentalis__ **Linnaeus.** American Sycamore, Planetree. Floodplains, alluvial islands, bottomlands, uplands. Panhandle, Escambia to Hamilton counties, sporadic in the northern peninsula; reported from Highlands and Lee counties, but probably not natural there.

Plumbaginaceae (Leadwort Family)

1. Flowers blue; leaves sessile or nearly so, the petiole-like leaf base more or less
   clasping the stem...…. ....*Plumbago auriculata*
1. Flowers white; leaves or some of them distinctly petiolate...........*Plumbago zeylanica*

*Plumbago auriculata **Lamarck.** Cape Leadwort. Disturbed sites, escaped from cultivation. Coastal counties of the central peninsula, Hillsborough, Lee, Brevard, and Martin
counties. South Africa.
Plumbago zeylanica **Linnaeus.** Doctorbush. Coastal hammocks, woodlands. Levy, Lake,
and Volusia counties southward along the coast; generally throughout the southernmost
peninsula, Lee and Martin counties southward to the Keys.

Poaceae (Grass Family)

1. Leaf veins forming distinctive square depressions, plants of northern and central
   Florida..…...….*Arundinaria gigantea*
1. Leaf veins not forming square depressions, plants of southern Florida
   2. Leaves pubescent..…...*Lasiacis ruscifolia*
   2. Leaves glabrous..…..…*Lasiacis divaricata*

Arundinaria gigantea **(Walter) Muhlenberg.** Switchcane. Moist slopes and uplands,
hammocks, stream and river banks, floodplains and bottoms. Panhandle and northern
peninsula, south to Polk, Osceola, and Brevard counties.
Lasiacis divaricata **(Linnaeus) Hitchcock.** Wild Bamboo, Smallcane, Florida Tibisee.
Subtropical hammock and hammock margins. Southern peninsula, Brevard and Lee coun-
ties southward, including the Keys.
*Lasiacis ruscifolia **(Kunth) Hitchcock & Chase.** Climbing Tibisee. Shell mounds.
Manatee County. Tropical America.

Podocarpaceae (Podocarpus Family)

__Podocarpus macrophyllus__ (Thunberg) **D. Don.** Yew Plumpine, Podocarpus. Disturbed sites, including upland woods and moist bottoms adjacent to lakes. Sporadically and sparingly naturalized from Escambia to Brevard and Sarasota counties; widely cultivated. China, Japan.

Polemoniaceae (Phlox Family)

__Phlox nivalis__ **Loddiges ex Sweet.** Trailing Phlox. Roadsides, sandhills. Panhandle, Escambia County, sporadically east and south to Duval, Sumter, and Lake counties; Manatee County.

Polygonaceae (Buckwheat Family)

1. Plant a climbing vine
 2. Many or most leaves heart-shaped at base, broadly ovate; flowers purple or rose......
 ....*Antigonon leptopus*
 2. Many or most leaves more or less truncate at base, narrowly ovate; flowers greenish white...*Brunnichia ovata*
1. Plant a tree or shrub
 3. Plant a large shrub or tree
  4. Leaves nearly circular, usually broader than long, base heart-shaped.................
  ....*Coccoloba uvifera*
  4. Leaves mostly ovate or broadly lanceolate, usually longer than broad, base wedge-shaped...*Coccoloba diversifolia*
 3. Plant a low-growing shrub, wiry and diffusely branched or erect, single-stemmed, and upright to about 1.5 m tall
  5. Collar of sheath at leaf nodes bearing short hairs..............*Polygonella robusta*
  5. Collar of sheath at leaf nodes lacking hairs
   6. Leaves 8–20 cm broad, plant usually single or very few stemmed, usually about 1 m tall...*Polygonella macrophylla*
   6. Leaves usually not exceeding about 6 mm broad, plant prostrate or low-growing, less than 0.5 m tall
    7. Plant more or less prostrate and mat forming, outer segments of the flower upright and appressed...................................*Polygonella myriophylla*

7. Plant upright, even if base of stem decumbent; outer segments of the flower spreading, reclining, or reflexed, not appressed
   8. Leaves 2–6 mm broad....................*Polygonella polygama* var. *polygama*
   8. Leaves not exceeding 1 mm broad...
   ....*Polygonella polygama* var. *brachystachya*

*◆*Antigonon leptopus* Hooker & Arnott. Coral Vine, Queen's Jewels. Hammock margins, disturbed sites, escaped from cultivation. Potentially statewide, including the Keys; more commonly naturalized in the central and southern peninsula. Mexico. FLEPPC listed (II).

***Brunnichia ovata* (Walter) Shinners.** Eardrop Vine, Ladies' Eardrops, American Buckwheatvine. River banks, floodplains, wet hammocks, especially in association with larger rivers. Panhandle, Escambia to Leon counties; Lafayette County.

***Coccoloba diversifolia* Jacquin.** Tie-tongue, Pigeon Plum, Dove Plum. Subtropical and coastal hammocks. Brevard and Lee counties southward, including the Keys.

***Coccoloba uvifera* (Linnaeus) Linnaeus.** Seagrape. Coastal hammocks, dunes, beach strands, disturbed sites, widely cultivated. Central and southern peninsula, Volusia and Pasco counties south along the coasts to the Keys.

!*Polygonella macrophylla* Small. Largeleaf Jointweed. State threatened. Beaches, stable dunes, sandy hammocks, scrub. Western panhandle, Escambia to Franklin counties. State threatened.

•!*Polygonella myriophylla* (Small) Horton. Small's Jointweed, Woody Wireweed, Sandlace. Scrub. Central peninsula, Orange, Polk, Osceola, and Highlands counties. Federally and state endangered.

***Polygonella polygama* (Ventenat) Engelmann & A. Gray.** October Flower. Beaches, sandhills, stable dunes, flatwoods, scrub. Western panhandle (Escambia to Wakulla counties), central and southern peninsula, south to Miami-Dade County, absent from the Keys.

•*Polygonella polygama* (Ventenat) Engelmann & A. Gray var. *brachystachya* (Meisner) Wunderlin. October Flower. Flatwoods. Southern peninsula, Osceola to Miami-Dade counties, absent from the Keys.

•*Polygonella robusta* (Small) G. L. Nesom & V. M. Bates. Largeflower Jointweed, Sandhill Wireweed. Sandhills, scrub. Central panhandle to south-central peninsula, Bay County east to Alachua County, south to Charlotte, Glades, and Palm Beach counties.

Proteaceae (Protea Family)

Grevillea robusta **A. Cunningham.** Silkoak. Disturbed sites, escaped from cultivation. Sporadically naturalized in the peninsula, Alachua to Lee and Broward counties. Australia.

Putranjivaceae (Putranjiva Family)

1. Sepals and stamens 4; fruit typically reddish- or orange-tinged at maturity, to about 1 cm long; leaves usually distinctly pointed at the apex..................*Drypetes lateriflora*
1. Sepals and stamens 5 or more; fruit typically whitish at maturity, to about 2 cm long; most or many leaves rounded at the aepx……...……..............*Drypetes diversifolia*

!*Drypetes diversifolia* **Krug & Urban.** Milkbark, Whitewood. Subtropical hammocks. Miami-Dade County and the Keys. State endangered.
!*Drypetes lateriflora* **(Swartz) Krug & Urban.** Guiana Plum. Subtropical hammocks. Southeastern coast, Brevard south to Collier and Miami-Dade counties and the Keys. State threatened.

Ranunculaceae (Buttercup Family)

1. Plant a low-growing, upright shrub, in Florida usually occurring on low banks of shallow, clear-water streams…......…...................….…*Xanthorhiza simplicissima*
1. Plant a vine
   2. Flowers solitary in the leaf axils, pinkish, bell-shaped, lower surface of some or many leaves distinctly glaucous beneath, at least when young………………………
   ………………………………………………………....*Clematis glaucophylla*
   2. Flowers in panicles or cymes, white, lower surface of leaf not glaucous
      3. Leaflets with entire margins; flowers bisexual…...................*Clematis terniflora*
      3. Leaflets usually toothed and/or lobed
         4. Leaves with 3 leaflets, these divided again into 3 parts, or leaves pinnately compound with 5 leaflets…......................…..........…..*Clematis catesbyana*
         4. Leaves with 3 undivided leaflets…......…….........…..……..*Clematis virginiana*

Clematis catesbyana **Pursh.** Virgin's-bower, Woodbine, Satincurls. Floodplains, margins of river swamps, riverbanks, hammocks, often in association with limestone. Central panhandle (Wathington to Leon counties), northwest peninsula (Dixie and Alachua to Polk counties), Duval County.

Clematis glaucophylla **Small.** Leather-flower, Whiteleaf Leather-flower. Rich slopes, floodplains, riverbanks, moist or wet hammocks, sometimes in association with limestone. Central panhandle (Jackson and Calhoun to Leon counties), Levy County.

*◆*Clematis terniflora* **de Candolle.** Sweet Autumn Virgin's-bower. Disturbed sites, lake margins, escaped from cultivation. Sporadically naturalized from Escambia to Baker counties, south to Hillsborough County. Japan. FLEPPC listed (II).

Clematis virginiana **Linnaeus.** Virgin's-bower, Woodbine. Disturbed sites, wet woods, streambanks, moist or wet hammocks. Sporadically distributed in the panhandle from Escambia to Alachua counties; west-central peninsula, Hernando to Hillsborough, Polk, and Hardee counties.

!*Xanthorhiza simplicissima* **Marshall.** Yellowroot. Margins of clear water streams, riverbanks. Sparsely distributed from Santa Rosa to Gadsden counties. State endangered.

Rhamnaceae (Buckthorn Family)

1. Leaves or many of them opposite or nearly so
 2. Leaf margins toothted; apex pointed........................…....*Sageretia minutiflora*
 2. Leaf margins entire; apex of many or most leaves notched
  3. Leaf blade stiff, usually widest at or above the middle; stipule usually one at the base of the petiole; fruit usually exceeding 1 cm long.......*Reynosia septentrionalis*
  3. Leaf blade pliable, usually widest at or below the middle; stipules usually paired at the base of the petiole; fruit not exceeding about 1 cm long.……..................
  ....*Krugiodendron ferreum*
1. Leaves clearly alternate
 4. Plant a vine (young plants sometimes appearing as a weakly erect, single stemmed herb or shrub)
  5. Margins entire; climbing by twining.......................…....*Berchemia scandens*
  5. Margins toothed, at least remotely so; climbing by tendrils.....*Gouania lupuloides*
 4. Plant a shrub or tree
  6. Plant with sharp-pointed stipular spines or axillary thorns
   7. Branchlets, inflorescences, and lower surface of the leaves densely hairy……...
   ....…......*Ziziphus mauritiana*
   7. Branchlets, inflorescences, and lower surface of leaves glabrous or nearly so, not densely hairy
    8. Leaves pinnately veined, the margins entire; blade not exceeding about 1 cm long.…..................................…..................…......…......*Ziziphus celata*
    8. Leaves with 3 major parallel veins, these arising from the base of the blade; margins conspicuously bluntly toothed; blade 3–7 cm long.....*Ziziphus jujuba*
  6. Plant lacking spines and thorns
   9. Plants of the central peninsula northward

10. Lateral leaf veins conspicuously regular, angling and ascending parallel to each other and curving slightly at the leaf margin; many leaves greater than 9 cm long (length varying 5–12 cm); plant becoming a tree..........................
....*Rhamnus caroliniana*
10. Lateral leaf veins not as above; all leaves less than 9 cm long; plant a shrub
   11. Leaves well under 1 cm long; branches yellow or yellowish green.........
   ....*Ceanothus microphyllus*
   11. Leaves well over 1 cm long; branches tan or brownish...........................
   ....*Ceanothus americanus*
9. Plants of the southern peninsula and Keys, predominantly Miami-Dade and Monroe counties, usually not occurring north of about Lee and Martin counties
   12. Leaf margins distinctly toothed; leaf blade with 3 veins arising from the base; plant a scandent shrub...*Colubrina asiatica*
   12. Leaf margins entire or obscurely bluntly toothed; mature plants forming a small tree or large shrub
      13. Apex of leaf rounded or blunt; veins conspicuously impressed on the upper surface...*Colubrina cubensis*
      13. Apex of leaf pointed; veins visible, but not conspicuously impressed on the upper surface
         14. Leaves lacking hairs or finely and inconspicuously hairy; leaf blades with 2–4 distinctive marginal glands at or near the base.................
         ....*Colubrina elliptica*
         14. Leaves distinctly and persistently finely reddish-hairy beneath (more generally reddish-hairy when young); leaf blade lacking marginal glands...*Colubrina arborescens*

Berchemia scandens (Hill) K. Koch. Rattan Vine, Alabama Supplejack, Supplejack. Moist upland woods, hammocks, swamps and swamp margins, wet woods, floodplains, bottomlands, wet flatwoods. Essentially throughout, except the Keys.

Ceanothus americanus Linnaeus. New Jersey Tea, Redroot. Sandhills, dry hammocks, longleaf pinelands, sandy roadsides. Panhandle and northern peninsula, Escambia to Duval counties, south to Volusia, Orange, and Highlands counties.

Ceanothus microphyllus Michaux. Littleleaf Buckbrush, Littleleaf Redroot. Sandhills, dry woods, flatwoods, well-drained pinelands. Panhandle and northern peninsula, Escambia to Nassau counties, south to Highlands County.

!Colubrina arborescens (Miller) Sargent. Greenheart, Coffee Colubrina, Snakebark. Subtropical Hammocks. Southern peninsula, Monroe and Miami-Dade counties and the Keys. State endangered.

*◆Colubrina asiatica (Linnaeus) Brongniart. Latherleaf, Asian Nakedwood, Colubrina. Disturbed sites, coastal hammocks, dunes, beaches, escaped from cultivation. Southern peninsula, Martin and Lee counties southward, including the Keys. Old World. FLEPPC listed (I).

!Colubrina cubensis (Jacquin) Brongniart var. floridana M. C. Johnston. Cuban Colubrina, Cuban Snakebark, Cuban Nakedwood. Subtropical hammocks, pinelands. Miami-

Dade County and the Keys. State endangered.

!*Colubrina elliptica* (**Swartz**) **Brizicky & W. L. Stern.** Soldierwood. Subtropical hammocks. Miami-Dade County and the Keys. State endangered.

Gouania lupuloides (**Linnaeus**) **Urban.** Chew Stick, Whiteroot. Coastal hammocks, margins of mangrove forests. More common in the southern peninsula (Miami-Dade County and the Keys), but scattered from Clay County southward along the coast; Manatee County.

Krugiodendron ferreum (**Vahl**) **Urban.** Black Ironwood, Leadwood. Hammocks, especially along the coast. Brevard County southward to Monroe County, including the Keys.

!*Reynosia septentrionalis* **Urban.** Darling Plum, Red Ironwood. Subtropical hammocks, margins of mangrove forests. Miami-Dade County and the Keys.

Rhamnus caroliniana **Walter.** Carolina Buckthorn. Moist deciduous forests, hammocks, shell middens, calcareous woodlands, usually in association with limestone. Irregularly distributed from the western panhandle to central peninsula, Escambia to St. Johns counties, south to Orange, Polk, and Sarasota counties.

Sageretia minutiflora (**Michaux**) **C. Mohr.** Buckthorn, Smallflower Mock Buckthorn. Calcareous woods and hammocks, sandy woodlands, shell mounds, floodplains. Panhandle south to southern peninsula, Escambia to Collier counties, more common in the western peninsula.

!*Ziziphus celata* **Judd & D. W. Hall.** Florida Ziziphus, Florida Jujube, Scrub Ziziphus. Scrub. Polk and Highlands counties. Federally and state endangered.

*Ziziphus jujuba **Miller.** Common Jujube. Disturbed sites, occasionally naturalized, escaped from cultivation. Central panhandle, Okaloosa to Gadsden counties. Europe, Asia.

*Ziziphus mauritiana **Lamarck.** Indian Jujube. Disturbed sites, escaped from cultivation. Broward and Miami-Dade counties. Southern Africa, China, Southeast Asia.

Rhizophoraceae (Red Mangrove Family)

1. Leaf blade usually flat, not reflexed upward from the midvein; leaf blade mucronate at apex; calyx lobes 4; flower petals entire..............................*Rhizophora mangle*
1. Leaf blade usually reflexed upward from the midvein; leaf blade not mucronate at apex; calyx lobes 5–16; petals 2-lobed....................................*Bruguiera gymnorrhiza*

*◆Bruguiera gymnorrhiza (**Linnaeus**) **Savigny.** Large-leaved Mangrove. Mangrove-dominated saltwater shorelines, escaped from cultivation. Miami-Dade County. Africa, Southeast Asia, Japan, Micronesia, Polynesia, subtropical Australia. FLEPPC listed (II).

Rhizophora mangle **Linnaeus.** Red Mangrove. Tidal swamps, saltwater shorelines, shallow waters of coastal bays, lagoons, creeks, and rivers. Coastal counties, Levy and Volusia counties southward, including the Keys, also collected in Wakulla County.

Rosaceae (Rose Family)

[**NOTE:** The key to the rose family includes a key to 17 species of hawthorns (*Cratae-gus*). There is substantial disagreement about the number of species of *Crataegus* that oc-cur in Florida, as well as the names that should be applied to them. Some experts estimate the number of species in Florida to exceed 40, others closer to 30, still others as few as 11. In deference to those workers who recognize more than the 17 species treated here, I have attempted to account for synonymous names by including them within parentheses within the key. Hence, several of the 17 species are keyed more than once, some several times. It is fair to say that not all *Crataegus* encountered in the field can be successfully keyed out below. Thus is the nature of this genus!]

1. Leaves compound
  2. Leaves with persistent, conspicuous stipules (at least when young) that are at least partially attached to the petiole
    3. Margins of the stipules deeply cut, dissected (comblike), or toothed
      4. Corolla with a single whorl of 5 white petals
        5. Branchlets glabrous or nearly so; flowers in many flowered panicles............ ....*Rosa multiflora*
        5. Branchlets with grayish matted hairs and longer gland-tipped hairs; flowers solitary or in few-flowered clusters.............................*Rosa bracteata*
      4. Corolla with numerous pink petals...............................*Rosa wichuraiana*
    3. Margins of the stipules not deeply cut or toothed, usually lined with glandular hairs
      6. Petals white; stipule attached to the petiole for less than half their length......... ....*Rosa laevigata*
      6. Petals pink; stipule attached to the petiole for most of their length
        7. Leaflets 3...*Rosa setigera*
        7. Leaflets of most leaves more than 3
          8. Plants of wetlands; major prickles recurved and hooked.......*Rosa palustris*
          8. Plants of well-drained upland woods; major prickles straight, diverging at right angles to the stem...*Rosa carolina*
  2. Leaves with inconspicuous, narrow stipules that are not attached to the petiole
    9. Stem erect, ascending, arching, scrambling, or clambering
      10. Petals distinctly rose-pink; leaves of flowering stems with 5–7 leaflets........... ....*Rubus niveus*
      10. Petals white, sometimes tinged pink; leaves of flowering stems predominantly with 3–5 leaflets
        11. Leaflets distinctly grayish-hairy and felty beneath; apex of leaflet usually rounded...*Rubus cuneifolius*
        11. Leaflets glabrous or hairy beneath, but not distinctly grayish and felty; apex of leaflet usually pointed............................*Rubus pensilvanicus*
    9. Stem prostrate or trailing

12. Stems with stout thorns and glandular hairs; branches of the inflorescence usually armed with prickles; flowers solitary.........................*Rubus trivialis*

12. Stems armed only with stout thorns, glandular hairs lacking; branches of the inflorescence usually unarmed; inflorescence usually 2–5-flowered..............
....*Rubus flagellaris*

1. Leaves simple

13. Pubescence of plant parts stellate; midvein and lowermost lateral veins arising together (or nearly so) from the base of the blade..............*Physocarpus opulifolius*

13. Pubescence of plant parts not stellate; venation pinnate, midvein and lowermost lateral veins not as above

14. Midveins and lower portion of larger lateral veins of upper leaf surface with at least a few dark purplish-red glands (seen with at least 10× magnification)........
....*Aronia arbutifolia*

14. Midveins and lower portion of larger lateral veins of upper leaf surface lacking glands

15. Main lateral veins appressed to the midvein for a short distance (<2 mm) before diverging (seen best from above with 10× magnification); flower petals narrowly oblong......................................*Amelanchier arborea*

15. Main lateral veins not appressed to the midvein below point of divergence; petals at least as broad as long

16. Ovary superior, visible within the floral tube; fruit a drupe

17. Flowers and fruits borne in racemes

18. Leaves deciduous, finely and evenly toothed, the tips of the teeth incurved

19. Leaf blade usually broadest near the middle; young branchlets and petioles glabrous; blade apex usually acute or acuminate...*Prunus serotina*

19. Leaf blade or many of them broadest above the middle; young branchlets and petioles hairy; apex of many blades rounded or blunt............................*Prunus alabamensis*

18. Leaves thick, evergreen, the margins entire or irregularly toothed

20. Leaf margins irregularly toothed...........*Prunus caroliniana*

20. Leaf margins entire............................*Prunus myrtifolia*

17. Flowers and fruits borne singly, or in fascicles or umbels

21. Flowers and fruit sessile or nearly so, stalk not exceeding about 2 mm long

22. Leaves at least 4 cm long; flowers deep pink; fruit fuzzy, to about 8 cm in diameter (the peach of commerce); branchlets not conspciuously zigzag.......................*Prunus persica*

22. Leaves not exceeding about 2.5 cm long; flowers white; fruit not fuzzy, to about 2.5 cm in diameter; branchlets conspciuously zigzag ....*Prunus geniculata*

21. Flowers and fruit distinctly stalked, the stalk usually exceeding 5 mm long

23. Bark shaggy, often curling in plates, brown, tan, or buff-colored; flowers about 2 cm or more across when fully open; apex of mature leaf blades usually conspicuously acuminate.. ....*Prunus americana*
23. Bark not shaggy; flowers usually not exceeding about 1.5 cm across when fully open; apex of mature leaf blades usually acute
    24. Marginal teeth of leaf blade tipped with a reddish or yellowish gland (these sometimes sloughing off with age); at least some blades strongly reflexed upward from the midvein.............................*Prunus angustifolia*
    24. Marginal teeth usually not tipped with a reddish or yellow gland; leaf blades usually flat....*Prunus umbellata*
16. Ovary inferior, wholly concealed within the floral tube; fruit a pome
  25. Branches usually thorny, the thorns sometimes leafy and terminating the branches
    26. Thorns usually terminating leafy branchlets
      27. Most leaves tending toward oblanceolate or spatulate, distinctly widest toward the tip, gradually and evenly tapering toward the base
        28. Leaf margins bluntly toothed, sometimes with only a few teeth evident..........................*Pyracantha fortuneana*
        28. Leaf margins entire...................*Pyracantha koidzumii*
      27. Most leaves elliptic, oblong, or ovate, widest near or below the middle.......................................*Malus angustifolia*
    26. Thorns naked, not bearing leaves or leaf scars, arising in the leaf axils
      29. Leaves conspicuously glandular on petiole and teeth, especially when young; twigs and branchlets conspicuously zigzag (geniculate)
        30. Leaf blades mostly less than 2 cm long or less than 1 cm broad
          31. Leaves often nearly as broad as long, broadly obovate or oval
            32. Leaves sharply toothed, often lobed; petiole tomentose; plant usually greater than 50 cm tall.. ....*Crataegus lassa* (*C. egens*)
            32. Leaves denticulate or remotely toothed, rarely lobed; petiole glabrate; plant usually less than 50 cm tall...*Crataegus lepida*
          31. Leaves 1.5–2 × longer than broad, spatulate or oblanceolate
            33. Leaves and pedicels glabrous.......................... ....*Crataegus lacrimata*

33. Leaves and pedicels pubescent, lanate, or tomentose
    34. Leaves generally spatulate, obscurely toothed to entire, blade attenuate along most of the petiole....………...*Crataegus lassa* (*C. recurva*)
    34. Leaves spatulate-oblanceolate, distinctly toothed, cuneate to petiole
        35. Thorns 1–3 cm; rarely >1 m tall.............……………...*Crataegus lepida* (*C. munda*)
        35. Thorns 3–6 cm; habit often >1 m tall.......…...…...............*Crataegus lepida* (*C. pexa*)
30. Leaf blades larger
    36. Leaves subentire along most of margin (*C. lassa* group, in part)
        37. Leaves mostly spatulate or short-obovate, tip obtuse or rounded, blade attenuate for most of the petiole.................*Crataegus lassa* (*C. recurva*)
        37. Leaves narrowly obovate or elliptic-obovate, tip pointed, blade cuneate or attenuate for about half of the petiole
            38. Leaves mostly elliptic-obovate, margins often subentire near apex...…........................*Crataegus lassa* (*C. integra*)
            38. Leaves mostly narrowly obovate or oblanceolate, margins often toothed near the apex..*Crataegus lassa*
    36. Leaves distinctly toothed
        39. Leaves often with 3 lobes or lobelike points at the apex; general leaf shape cuneate or narrowly obovate
            40. Fruit small, usually <10 mm diameter, often with calyx elevated; leaf base often attenuate or long-cuneate (*C. lassa* group, in part)
                41. Leaf apex flattened, rounded or obtuse; marginal lobes may be obscure...................................*Crataegus lassa* (*C. vicana*)
                41. Leaf apex obtuse to acute; marginal lobes usually conspicuous
                    42. Leaf lobes acute, distally projecting, serrations conspicuous; blades to 3 cm long.....................................…........……...…*Crataegus lassa* (*C. quaesita*)
                    42. Leaf lobes blunt to acute, largest ones somewhat divergent, serrations often

inconspicuous (crenate or crenulate)

43. Petioles 10–20 mm long.............

.......*Crataegus lassa* (*C. floridana*)

43. Petioles 3–6 mm long...............

.........*Crataegus lassa* (*C. furtiva*)

40. Fruit often 10–12 mm, calyx sessile; leaf base cuneate

44. Thorns abundant, present at nearly every node, averaging 3–6 cm long................

........................*Crataegus lepida* (*C. pexa*)

44. Thorns less numerous or shorter

45. Lobes usually obtuse, leaves often 3–5 cm long......*Crataegus lassa* (*C. lanata*)

45. Lobes usually acute, leaves 2–4 cm......

...........*Crataegus lassa* (*C. meridiana*)

39. Leaves unlobed, or with more than 3 lobelike points, these beginning near middle of blade; general leaf shape broadly obovate, ovate, or oval

46. Leaves often 5–8 cm long; fruit yellow, elongate ....*Crataegus flava*

46. Leaves <5 cm; fruit usually orange to red, subglobose

47. Stamens 10; leaves widest at midpoint of blade; branch tips of crown mostly ascending and crooked......................

..................................*Crataegus aprica*

47. Stamens 20; leaves often widest beyond midpoint of blade; branch tips of crown recurved or drooping

48. Leaf teeth small; apex acute and margin often with 1 or 2 lobelike points each side (*C. aprica* group)

49. Anthers usually white or pink; pedicels tomentose; terminal shoot leaves often orbicular

50. Pedicels tomentose..............

..*Crataegus aprica* (*C. sororia*)

50. Pedicels sparsely hairy..........

Crataegus aprica (*C. leonensis*)

49. Anthers usually purplish; pedicels pubescent or glabrate; terminal shoot leaves mostly broadly elliptic or obovate

51. Leaves mostly unlobed, obovate
   52. Marginal teeth crenate; blades often slightly pubescent abaxially.......... *Crataegus aprica* (*C. visenda*)
   52. Marginal teeth sharp; blades glabrous abaxially at maturity.......................... *Crataegus aprica* (*C. galbana*)
51. Leaves often with 1 or 2 pairs of blunt lobes; blades broadly obovate or rhombic............... ...*Crataegus aprica* (*C. egregia*)
48. Leaf teeth sharp, distinct; apex rounded to abruptly short-pointed, margin rarely with lobelike points ....(*C. alabamensis* group)
   53. Leaves and inflorescence parts densely pubescent to tomentose; leaf margins finely crenate-serrate
      54. Leaves pubescent, apex mostly rounded.................................... ....*Crataegus alabamensis*
      54. Leaves pubescent when young, later glabrate, apex with a short point... *Crataegus alabamensis* (*C. ravenelii*)
   53. Leaves and inflorescence thinly pubescent to glabrate; leaf margin dentate... *Crataegus alabamensis* (*C. florens*)
29. Leaves eglandular, or if glandular then twigs relatively straight, not conspicuously zigzag
   55. Primary lateral veins of lobed leaves terminating at the base of the sinuses, as well at the tips of the lobes
      56. Leaves spatulate or oblanceolate, most less than 1 cm wide, cuneate at base, often with 3 terminal lobes....... ....*Crataegus spathulata*
      56. Leaves deltoid or broadly ovate, greater than 2 cm wide, truncate or slightly cordate at base
         57. Leaves thin, dull, deeply laciniate, teeth acuminate; petiole hairy.......*Crataegus marshallii*
         57. Leaves subcoriaceous, glossy, teeth of lobes acute

    or blunt; petiole glabrous.......................................
    ....*Crataegus phaenopyrum*
55. Primary lateral veins of lobed leaves terminating only in
    the tips of the lobes
    58. Leaves with hair tufts in abaxial main vein axil;
        [plants typically of wet or floodplain habitats]
        59. Inflorescence simple, 1- to 5-flowered; fruit
            greater than 1 cm diameter, maturing late spring
            [*C. aestivalis* group]
            60. Leaf blades mostly 5–7 cm long, elliptic........
            ....*Crataegus opaca*
            60. Leaf blades mostly less than 5 cm long, many
                obovate
                61. Upper surface of leaf usually glossy, lower
                    surface with tufts of whitish hairs in the
                    vein axils...............*Crataegus aestivalis*
                61. Upper surface of leaf dull, lower surface
                    with tufts of reddish hairs....................
                ....*Crataegus rufula*
        59. Inflorescence compound, 5- to 20-flowered; fruit
            usually less than 1 cm in diameter, maturing in
            autumn
            62. Petiole 5–12 mm long; terminal shoot leaves
                rarely lobed..
                ......*Crataegus crus-galli* var. *pyracanthifolia*
            62. Petiole greater than 15 mm long; terminal
                shoot leaves rarely unlobed...*Crataegus viridis*
    58. Leaves glabrous or with hairs scattered, not in tufts;
        [plants typically of upland habitats]
        63. Leaves mostly < 3 cm long; calyx lobes
            conspicuously foliaceous, deeply toothed; spines
            slender...............................*Crataegus uniflora*
        63. Leaves commonly greater than 3 cm long and not
            with above combination of characters
            64. Petiole eglandular
                65. Leaves glabrous, mostly broadly elliptic or
                  obovate.....................*Crataegus crus-galli*
                65. Leaves often hairy along the veins on the
                    lower surface; mostly oblanceolate............
                  ..*Crataegus crus-galli* var. *pyracanthifolia*
            64. Petiole glandular
                66. Leaf serrations small, crenate or crenate-
                  serrate; leaves barely lobed; stamens 10;

leaf apex acute; fruit elongate...............
..................................*Crataegus flava*
66. Leaf serrations sharp or coarse; leaves
usually lobed; usually with dark glands at
the tip of marginal teeth (*C. pulcherrima*
group)
67. Leaves ovate-lanceolate, lobe tips
forming a straight line from about
midblade to leaf apex
68. Terminal shoot leaves usually with
4 or 5 pairs of distinct lobes; plant
often a suckering shrub..................
.*Crataegus pulcherrima* (*C. incilis*)
68. Terminal shoot leaves with 3 to 4
pairs of obscure to moderate
lobes; plant usually arborescent....
...*Crataegus pulcherrima* (*C. gilva*)
67. Leaves elliptic, ovate or widely ovate,
lobe tips usually not forming a straight
line from midpoint to leaf apex
69. Leaf blades of terminal shoots
longer than broad, lobes blunt to
acute..........*Crataegus pulcherrima*
69. Leaf blades of terminal shoots
often as broad as long, lobes often
rounded.................................
.*Crataegus pulcherrima* (*C. opima*)
25. Branches rarely thorny, sometimes with sharp-tipped branchlets
70. Many or most leaves 12–25 cm long; leaf margins coarsely
toothed; veins on upper surface of blade conspicuous;
inflorescence densely hairy.........................*Eriobotrya japonica*
70. Many or most leaves less than 12 cm long; leaf margins finely
and often bluntly toothed; veins on upper surface comparatively
obscure; inflorescence glabrous or sparsely hairy
71. Flowers pink; leaf apex blunt................*Malus angustifolia*
71. Flowers white; leaf apex acuminate
72. Flowers mostly 2.5–3 cm across when fully open, styles
3; fruit bronze-colored, roughened.........*Pyrus calleyrana*
72. Flowers mostly 4 cm across when fully open, styles 5;
fruit yellow green, smooth (the pear of commerce).........
....*Pyrus communis*

Amelanchier arborea (**Michaux f.**) **Fernald.** Serviceberry, Downy Serviceberry, Common Serviceberry, Sarvisberry, Shadbush. Streambanks, open woods, mesic forests, dry slopes and hammocks. Panhandle, Escambia to Leon counties.

Aronia arbutifolia (**Linnaeus**) **Persoon.** Red Chokeberrry. Moist flatwoods, bog margins, margins of titi swamps and wet savannas, spring runs. Panhandle and northern peninsula, south to Manatee, Hardee, Highlands, Osceola, and Brevard counties.

Crataegus aestivalis (**Walter**) **Torrey & A. Gray.** May Haw, May hawthorn. Series Aestivales. Edges of wetland depressions, floodplains, swamp margins, typically in and near standing water; throughout the panhandle and southward to about Levy and Volusia counties.

Crataegus alabamensis **Beadle.** Alabama Hawthorn. Series Lacrimatae. Includes *C. florens, C. ravenelii*. Upland pine and pine–oak forests, mostly in xeric or nearly xeric habitats with sandy or well-drained clay soils; rare in northern Florida, mostly near Tallahassee.

Crataegus aprica **Beadle.** Sunny Hawthorn. Series Apricae. Includes *C. egregia, C. galbana, C. leonensis, C. sororia, C. visenda*. Distributed in Florida from about the central panhandle to the northern peninsula, south to about Alachua County.

Crataegus crus-galli **Linnaeus.** Cockspur Hawthorn. Series Crus-galli. Open woodlands and upland woods; throughout northern Florida and the eastern panhandle from about Bay and Jackson to Sumter, Lake, and Volusia counties.

Crataegus crus-galli var. *pyracanthifolia* **Aiton.** Narrowleaf Cockspur. Series Crus-galli. Hammocks and open woodlands, Leon to Madison counties.

Crataegus flava **Aiton.** Yellow Hawthorn. Series Intricatae. Dry woods; very rare if extant in Florida; Gadsden County.

Crataegus lacrimata **Small.** Weeping Hawthorn, Pensacola Hawthorn. Series Lacrimatae. Xeric habitats, typically in deep sandy soils; as treated here confined to the western and central panhandle from about Escambia to Calhoun counties.

Crataegus lassa **Beadle.** Sandhill Hawthorn. Series Lacrimatae. Includes *C. egens, C. floridana, C. furtiva, C. integra, C. lanata, C. meridiana, C. quaesita, C. recurva, C. vicana*. A variable species distinguished by its mostly narrowly oblanceolate or narrowly obovate leaves, mostly entire leaf margins (except near the apex), and often flat or bluntly rounded apex.

Crataegus lepida **Beadle.** Dwarf Hawthorn. Series Lacrimatae. Includes *C. munda, C. pexa*.

Crataegus marshallii **Eggleston.** Parsley Hawthorn. Series Apiifoliae. Wooded slopes, moist woods, floodplains; throughout northern Florida, southward to about Polk and Hillsborough counties.

Crataegus opaca **Hooker & Arnott.** Western May Haw. Series Aestivales. Occurring in similar places as eastern mayhaw; rare in Florida, essentially restricted to the westernmost panhandle, Santa Rosa and Escambia counties; more common west of Mississippi's Pearl River.

!*Crataegus phaenopyrum* (**Linnaeus f.**) **Medikus.** Washington Hawthorn. Series Cordatae. Low woods; known or reported in Florida only in the panhandle, including Walton, Washington, Liberty, and Wakulla counties; easily overlooked and potentially more widespread than currently known. State endangered.

Crataegus pulcherrima **Ashe.** Beautiful Hawthorn. Series Pulcherrimae. Includes *C.*

gilva, C. incilis, C. opima. Open upland woods; distributed in Florida from Walton to Jefferson counties, and from northeast Florida southward to Alachua County.

***Crataegus rufula* Sargent.** Rufous May Haw. Series Aestivales. Habitats similar to those of eastern mayhaw; restricted in Florida to the central panhandle, especially eastern Jackson County.

***Crataegus spathulata* Michaux.** Littlehip Hawthorn. Series Microcarpae. Bottomlands, floodplains, and wooded slopes, often where calcareous; limited in Florida to the central panhandle from about Holmes to Gadsden and Liberty counties.

Crataegus uniflora* *Münchhausen. Oneflower Hawthorn. Series Parvifoliae. Open sandy woods, roadsides; Washington County to northeastern Florida, south to Polk County.

***Crataegus viridis* Linnaeus.** Green Hawthorn. Series Virides. Low woods, pond edges, swamps; across northern Florida, south to about Levy, Marion, and Volusia counties.

**Eriobotrya japonica* (Thunberg) Lindley. Loquat, Japanese Plum. Disturbed sites, hammocks, escaped from cultivation. Naturalized in scattered locations throughout the state, except the Keys. China. Potentially invasive.

!*Malus angustifolia* (Aiton) Michaux. Southern Crabapple. Open, upland woods and pinelands, hammocks, calcareous woodlands. Central panhandle, Washington to Taylor and Hamilton counties. State threatened.

!*Physocarpus opulifolius* (Linnaeus) Maximowicz. Common Ninebark, Ninebark. Stream margins, open disturbed sites. Jackson and Calhoun counties. State endangered.

***Prunus alabamensis* C. Mohr.** Alabama Cherry. Sandy pine—oak woodlands. Panhandle, Escambia to Calhoun and Liberty counties.

***Prunus americana* Marshall.** American Plum. Upland woods, mesic hammocks, elevated sites within floodplains, disturbed sites, often in association with calcareous soils. Scattered locations in the central panhandle and northern peninsula, Jackson to Nassau counties, south to Citrus and Lake counties.

***Prunus angustifolia* Marshall.** Chickasaw Plum. Woodland margins, fencelines, roadsides, hedge rows, sandy pinelands. Panhandle and northern peninsula, Escambia County east to Duval County, south to Lake, Pinellas, and Hillsborough counties.

***Prunus caroliniana* (Miller) Aiton.** Carolina Laurelcherry. Widely established in weedy sites, along roadsides, suburban landscapes, hammocks, upland woods. Native to Florida but often cultivated and now widespread beyond what would likely have been its natural habitat. Panhandle and northern peninsula, generally south to Sarasota, DeSoto, Okeechobee, and St. Lucie counties, also reported from Broward County.

•!*Prunus geniculata* R. M. Harper. Scrub Plum. Scrub, roadsides adjacent to scrub. Central peninsula, Lake to Highlands counties. State endangered.

!*Prunus myrtifolia* (Linnaeus) Urban. West Indian Cherry. Rockland hammocks. Miami-Dade County. State threatened.

*Prunus persica (Linnaeus) Batsch. Peach. Disturbed sites, margins of fields, persisting near old home sites, escaped from cultivation. Sporadically naturalized, panhandle and northern peninsula, south to Hernando, Lake, and Brevard counties. China.

***Prunus serotina* Ehrhart.** Black Cherry. Upland woods, hammocks, disturbed sites, roadsides, suburban landscapes, powerline easements, fencerows, weedy and spreading into open woodlands, often by birds. Panhandle and northern peninsula, south to DeSoto,

Manatee, and Polk counties.

Prunus umbellata **Elliott.** Flatwoods Plum, Hog Plum. Hammocks, flatwoods, mixed woods. Panhandle and northern peninsula, south to Sarasota, DeSoto, Highlands, and Martin counties.

*Pyracantha fortuneana (**Maximowicz**) **H. L. Li.** Chinese Firethorn. Disturbed sites, escaped from cultivation. Okaloosa County. China.

*Pyracantha koidzumii (**Hayata**) **Rehder.** Formosa Firethorn. Disturbed sites, woodlands, margins of limestone glades, escaped from cultivation. Sporadically naturalized in the panhandle and northern peninsula, south to Volusia and Hillsborough counties. Taiwan. Potentially invasive.

*Pyrus calleryana **Decaisne.** Callery Pear. Disturbed sites, old fields, roadsides, escaped from cultivation. Panhandle, east to Dixie and Suwannee counties. China. Potentially invasive.

*Pyrus communis **Linnaeus.** Common Pear. Disturbed sites, fencerows, thickets, clearings, often near human dwellings, escaped from cultivation. Sporadically and potentially naturalized, panhandle and northern peninsula, south to Brevard County. Europe.

*Rosa bracteata **J. C. Wendland.** McCartney Rose. Disturbed sites, roadsides, hammocks, escaped from cultivation. Sporadically naturalized, panhandle south to Citrus, Marion, and Volusia counties. China.

Rosa carolina **Linnaeus.** Carolina Rose. Mixed upland woods, pine—oak forests, hammocks. Central panhandle (Jackson to Liberty and Leon counties), northern peninsula (Alachua and Clay counties).

*Rosa laevigata **Michaux.** Cherokee Rose. Roadsides, disturbed woods, hammocks, escaped from cultivation. Panhandle (Santa Rosa to Jefferson counties), central peninsula (Alachua and Volusia to Manatee and Highlands counties). China. Potentially invasive.

*Rosa multiflora **Thunberg ex Murray.** Multiflora Rose. Disturbed sites, mostly near plantings, escaped from cultivation. Panhandle, Escambia to Leon counties. Asia. Potentially invasive.

Rosa palustris **Marshall.** Swamp Rose. Banks of streams and spring runs, margins of ponds and wetlands, marshes, swamps. Central panhandle and northern peninsula, Jackson, Liberty, and Franklin counties east to Duval County, south to Pinellas, Hillsborough, and Polk counties.

Rosa setigera **Michaux.** Climbing Rose. Stream banks. Ostensibly the panhandle, based on a single collection by A. W. Chapman that was not attributed to county.

*Rosa wichuraiana **Crépin ex Déséglise.** Memorial Rose. Dry, disturbed sites, roadsides, railroad banks. Sporadically naturalized in the panhandle, Escambia to Gulf counties; also naturalized in Hardee County. Asia.

Rubus cuneifolius **Pursh.** Sand Blackberry. Sandy woods, dry disturbed sites, sandy pinelands, flatwoods. Panhandle and peninsula, south to Charlotte, Glades, and Broward counties.

Rubus flagellaris **Willdenow.** Northern Dewberry. Hammocks, sandhills, pinelands, lowland woods, stream banks, old fields, margins of second growth uplands. Panhandle, Santa Rosa to Gadsden, Liberty, and Franklin counties.

*Rubus niveus **Thunberg.** Snowpeaks Raspberry. Disturbed sites, escaped from cultiva-

tion. Miami-Dade County. Asia.

***Rubus pensilvanicus* Poiret.** Sawtooth Blackberry, Pennsylvanica Blackberry. Wet woodlands, streambanks, often near standing water but also in fields, roadsides, upland woods, and roadsides. Panhandle and northern peninsula, south to Charlotte, Glades, and Okeechobee counties.

***Rubus trivialis* Michaux.** Southern Dewberry. Well-drained and poorly drained woods, disturbed sites, roadsides, old fields, margins of marshes, suburban landscapes, hedge- and fencerows. Panhandle and peninsula, south to Collier and Broward counties.

Rubiaceae (Madder Family)

1. Plant a vine, vinelike, or a scrambling or reclining shrub
    2. Stem strictly prostrate, creeping and rooting at the nodes; many leaves nearly circular in outline; flowers paired, axillary; fruit red....................*Mitchella repens*
    2. Plants lacking the above combination of characters
        3. Plant mostly a climbing, twining vine; flowers lilac in color; fruit yellowish-brown or orange-brown
            4. Fruit compressed or flattened; seeds conspicuously winged........................
            ....…........*Paederia cruddasiana*
            4. Fruit rounded; seeds not winged...............................…......*Paederia foetida*
        3. Plant mostly a scrambling or reclining shrub, sometimes with elongated, vinelike branches; flowers reddish, pinkish, white, or yellowish white; fruit yellow or white
            5. Flowers and fruit sessile or nearly so; leaves linear or narrowly lanceolate; flowers usually pink or reddish exteriorily (sometimes white); fruit yellow
                6. Leaves with only a single prominent midvein...................*Ernodea cokeri*
                6. Leaves with 3 or more prominent longitudinal veins…........*Ernodea littoralis*
            5. Flowers and fruit stalked; leaves elliptic; flowers white; fruit white
                7. Plant low-growing, often trailing...................…..........*Chiococca parvifolia*
                7. Plant a large, sprawling, vineline shrub...........................*Chiococca alba*
1. Plant an erect shrub or tree
    8. Flowers and fruit borne in a dense, rounded, headlike cluster
        9. Fruit cluster brownish, "prickly," dry.....................*Cephalanthus occidentalis*
        9. Fruit cluster greenish when young, becoming yellow, not "prickly," fleshy
            10. Blades of larger leaves not exceeding about 10 cm long...........*Morinda royoc*
            10. Blades of larger leaves at least 20 cm long....................…......*Morinda citrifolia*
    8. Flowers and fruit not borne in dense, rounded, headlike clusters
        11. Plants thorny
            12. Leaves 2–5 cm long; corolla 5-lobed; fruit 6–13 mm long....*Randia aculeata*
            12. Leaves not exceeding about 1 cm long; corolla 4-lobed; fruit 2–4 mm long...
            ....*Catesbaea parviflora*

11. Plants not thorny
>13. At least some flowers on any plant with at least one much enlarged, petal-like, pinkish, whitish, or yellowish ovate sepal 6–7 cm long; very young twigs densely hairy; veins on upper surface of leaf raised.....................
....*Pinckneya bracteata*
>13. Plants lacking the above combination of characters
>>14. Flowers or fruits axillary, solitary or cymose
>>>15. Upper surface of leaves harshly scabrous, or at least moderately or sparsely hairy on one or both surfaces
>>>>16. Upper surface harshly scabrous; blade stiff, usually broadest above the middle..*Guettarda scabra*
>>>>16. Upper surface finely scabrous or smooth; blade more or less pliable, usually broadest at or below the middle.....................
....*Guettarda elliptica*
>>>15. Leaf surfaces glabrous
>>>>17. Leaf blade usually widest above the middle, the apex rounded or notched; fruit a berry, 5–8 cm long.................*Genipa clusiifolia*
>>>>17. Leaf blade usually oblong or ovate, usually widest at or below the middle; fruit a woody capsule, 1–1.5 cm long........................
....*Exostema caribaeum*
>>14. Flowers or fruits in terminal compound cymes
>>>18. Leaves with a single prominent midvein, the lateral veins obscure, inconspicuous
>>>>19. Low, profusely branched shrub usually not exceeding about 1 m tall; leaves linear or narrowly lanceolate, hairy; leaf margins revolute...*Strumpfia maritima*
>>>>19. Densely foliaged shrub, usually exceeding 1 m tall at maturity (potentially to at least 3 m tall); leaves elliptic, ovate, or nearly circular, glabrous; leaf margins not revolute......*Erithalis fruticosa*
>>>18. Lateral veins prominent, conspicuous
>>>>20. Corolla finely hairy on the outer surface
>>>>>21. Flowers purple, white, or rosy throughout, tips of the corolla lobes conspicuous, spreading into 5 distinct lobes; fruit a capsule.......................................*Pentas lanceolata*
>>>>>21. Flowers orange or reddish-orange outside, yellowish inside, tips of the petal lobes usually erect, barely if at all spreading; fruit a red or purplish berry......................*Hamelia patens*
>>>>20. Corolla glabrous on the outer surface
>>>>>22. Tube of the corolla exceedingly slender
>>>>>>23. Leaves with a short petiole; flowers white, corolla tube not exceeding about 1 cm long....................*Ixora arborea*
>>>>>>23. Leaves sessile; flowers red, corolla tube usually well over 1 cm long (to about 5 cm)....................*Ixora coccinea*
>>>>>22. Tube of the corolla wider, the flower in the shape of a funnel

24. Leaf blades with punctate dots (seen best when held
against transmitted light)..................*Psychotria punctata*
24. Leaf blades lacking punctate dots
   25. Upper surface of leaf lustrous green; calyx lobes
   reduced to minute teeth...............*Pyschotria nervosa*
   25. Upper surface of leaf dull; calyx conspicuously lobed
     26. Leaf glabrous; most leaf blades not exceeding
     about 6 cm long..............*Psychotria ligustrifolia*
     26. Leaf densely and finely hairy; most blades greater
     than 10 cm long..................*Psychotria sulzneri*

!*Catesbaea parviflora* **Swartz.** Lilythorn, Smallflower Lilythorn, Dune Lilythorn. Pine rocklands, sandy woods, stable back dunes. Keys. State endangered.

Cephalanthus occidentalis **Linnaeus.** Buttonbush, Honey-balls, Globe-flower. Swamps, pond and lake margins, pineland depressions, sloughs, streambanks, usually where water stands much of the time. Essentially throughout, except the Keys.

Chiococca alba **(Linnaeus) Hitchcock.** Snowberry, Milkberry. Hammocks and hammock margins, pinelands, shell middens, often in association with limestone. Coastal counties. Dixie and Duval counties southward, including the Keys.

Chiococca parvifolia **Wullschlägel ex Grisebach.** Snowberry. Pinelands, margins of rockland hammocks. Southernmost peninsula and the Keys.

!*Erithalis fruticosa* **Linnaeus.** Blacktorch. Coastal scrub, coastal hammocks, dunes. Southeast coast, Martin County south to the Keys. State threatened.

!*Ernodea cokeri* **Britton ex Coker.** Coker's Beach Creeper, One-nerved Ernodea. Hammocks, pinelands. Miami-Dade County and the Keys. State endangered.

Ernodea littoralis **Swartz.** Beach Creeper, Golden Creeper, Coughbush. Coastal dunes. Coastal counties, Pinellas and Volusia counties southward, including the Keys.

!*Exostema caribaeum* **(Jacquin) Roemer & Schultes.** Princewood, Caribbean Princewood. Hammocks, pinelands. Miami-Dade County and the Keys. State endangered.

Genipa clusiifolia **(Jacquin) Grisebach.** Seven-Year Apple. Coastal hammocks and adjacent transition zone. Southern peninsula, Lee and Broward counties southward, including the Keys.

Guettarda elliptica **Swartz.** Velvetseed, Everglades Velvetseed, Hammock Velvetseed. Hammocks, pinelands. Southern peninsula, St. Lucie and Broward counties, southward, including the Keys.

Guettarda scabra **(Linnaeus) Ventenat.** Rough Velvetseed. Pinelands, hammocks. Southeastern coast, Martin County southward, including the Keys.

Hamelia patens **Jacquin.** Firebush. Hammocks, roadsides, disturbed sites. Central peninsula, Marion, Volusia, and Pasco counties southward; potentially escaped from plantings and adventive farther north.

*Ixora arborea **Roxburgh ex J. E. Smith.** Smallflower Jungleflame. Hammocks, disturbed sites, escaped from cultivation. Sporadically naturalized in the southern peninsula, Palm Beach to Miami-Dade counties. Asia and Africa.

Ixora coccinea **Linnaeus.** Scarlet Jungleflame. Disturbed sites, escaped from cultivation. Southern peninsula, Manatee, Lee, Collier, Palm Beach, and Broward counties. India.

Mitchella repens **Linnaeus.** Partridgeberry, Twinberry. Hammocks, rich upland woods. Panhandle and peninsula, south to Sarasota, DeSoto, Highlands, and Martin counties.

*Morinda citrifolia **Linnaeus.** Indian Mulberry. Mangrove margins, hammocks, escaped from cultivation. Southern peninsula, Miami-Dade County and the Keys. India.

Morinda royoc **Linnaeus.** Redgal, Morinda, Yellow Root, Cheese Shrub, Mouse's Pineapple. Coastal and subtropical hammocks. Hillsborough and Brevard counties southward, including the Keys.

*◆Paederia cruddasiana **Prain.** Sewervine. Disturbed sites, hammock margins, escaped from cultivation. Miami-Dade County. Southern Asia. FLEPPC listed (II).

*◆Paederia foetida **Linnaeus.** Skunkvine. Hammocks, upland woods, disturbed sites, escaped from cultivation. Central panhandle, Gadsden and Franklin counties southward to Broward County. Asia. FLEPPC listed (I).

*Pentas lanceolata **(Forsskål) Deflers.** Egyptian Starcluster. Disturbed sites, escaped from cultivation. Miami-Dade County. Yemen.

!Pinckneya bracteata **(W. Bartram) Rafinesque.** Fevertree, Pinckneya. Creek swamps, drainages, bog margins, wetland creek heads in pinelands. Central panhandle (Jackson, Washington, and Bay counties east to Jefferson County), northern peninsula (Marion and Clay counties). State threatened.

!Psychotria ligustrifolia **(Northrop) Millspaugh.** Bahama Wild Coffee, Wild Coffee. Rockland hammocks, pinelands. Miami-Dade County and the Keys. State endangered.

Psychotria nervosa **Swartz.** Wild Coffee. Hammocks, pinelands, disturbed woodland margins. Peninsula, Duval, Alachua, and Levy counties southward, including the Keys.

*Psychotria punctata **Vatke.** Wild Coffee, Dotted Wild Coffee. Disturbed sites, escaped from cultivation. Keys. Africa.

Psychotria sulzneri **Small.** Wild Coffee, Shortleaf Wild Coffee. Hammocks. Peninsula, Citrus, Sumter, Lake, and Volusia counties southward, except the Keys.

Randia aculeata **Linnaeus.** Randia, Indigo Berry, White Indigo Berry. Coastal hammocks, woodland margins. Southwestern and southeastern coasts, Hillsborough and Brevard counties southward, including the Keys.

!Strumpfia maritima **Jacquin.** Pride-of-Big-Pine, Strumpfia, Snowbank. Coastal strand, wet pinelands, often in more or less saline soils. Keys. State endangered.

Rutaceae (Citrus Family)

1. All leaves with a single leaflet, appearing simple
  2. Fruit predominantly 15 cm long or more (some fruit may be shorter than this).........
  ....*Citrus medica*
  2. Fruit predominantly less than 15 cm long
    3. Petiole usually or often winged; fruit yellow, orange, or green; flowers with 20 stamens
      4. Fruit usually not exceeding about 4 cm in diameter.............…...*Citrus japonica*
      4. Fruit usually 5 cm or more in diameter.........................…..…....*Citrus reticulata*
    3. Petiole not winged; fruit black, less than 1 cm diameter; flowers with 10 stamens..
    ....….........…...*Severinia buxifolia*
1. Most or many leaves with 3 or more leaflets, distinctly compound
  5. Branches and sometimes the leaves armed with thorns or prickles
    6. Leaflets 5–9
      7. Leaves odd-pinnate, with a distinct terminal leaflet
        8. Petiole and rachis winged...*Zanthoxylum fagara*
        8. Petiole and rachis not winged
          9. Plant a shrub; prickles usually borne in pairs at the leaf nodes; leaves usually lacking prickles, the leaflets usually at least sparsely hairy on the veins beneath.....….........................……...........*Zanthoxylum americanum*
          9. Plant a tree; prickles usually scattered along the stem between the leaf nodes; leaves often bearing prickles; surfaces of the leaflets glabrous.......
          ....….......................................…....*Zanthoxylum clava-herculis*
      7. Leaves even-pinnate, lacking a terminal leaflet..........*Zanthoxylum coriaceum*
    6. Leaflets usually 3 (sometimes 1 or 2)
      10. Petiole winged; fruit resembling a small orange, 4–5 cm diamter................
      ....*Poncirus trifoliata*
      10. Petiole not winged; fruit berrylike, to 1.5 cm diameter.........*Triphasia trifolia*
  5. Branches and leaves not armed
    11. Leaflets on most or many leaves 5 or fewer
      12. Fruit a dry, circular, wafer-like samara.........................*Ptelea trifoliata*
      12. Fruit fleshy, a drupe or berry
        13. Fruit a blue or black drupe; leaflets usually not exceeding about 7 cm long
          14. Inflorescence finely hairy.........................…*Amyris balsamifera*
          14. Inflorescence glabrous.........…....................…....*Amyris elemifera*
        13. Fruit a white or pink berry; few to many leaflets exceeding 7 cm long.....
        ....*Glycosmis parviflora*
    11. Leaflets on many leaves more than 5
      15. Lateral leaflets all opposite................................*Zanthoxylum flavum*
      15. Lateral leaflets alternate, opposite, and sub-opposite......*Murraya paniculata*

Amyris balsamifera **Linnaeus.** Balsam Torchwood. Subtropical Hammocks. Miami-Dade County and the Keys.

Amyris elemifera **Linnaeus.** Torchwood, Sea Torchwood. Wet or moist coastal hammocks. East coast, Flagler County to the Keys.

*Citrus japonica **Thunberg.** Kumquat. Disturbed sites, rarely escaped from cultivation. Glades County. China.

*Citrus medica **Linnaeus.** Citron. Escaped from cultivation and sparsely naturalized. Panhandle and southern peninsula. Northern India.

Citrus reticulata* **Blanco. Tangerine. Widely planted, escaped from cultivation and sporadically naturalized. Central peninsula. Southeast Asia, China.

Citrus hybrids. Not keyed.

    Citrus × *aurantiifolia* **(Christmann) Swingle.** Key Lime.

    Citrus × *aurantium* **Linnaeus.** Sour Orange, Grapefruit, Sweet Orange.

    Citrus × *jambhiri* **Lush.** Mandarin Lime, Rough Lemon.

    Citrus × *limon* **(Linnaeus) Burman f.** Lemon.

Glycosmis parviflora* **(Sims) Little. Flower Axistree. Disturbed sites, rockland hammocks, escaped from cultivation. Broward County south to the Keys. China, Japan.

*◆*Murraya paniculata* **(Linnaeus) Jack.** Orange Jessamine. Disturbed sites, hammocks. Broward County south to the Keys. Tropical Asia. FLEPPC listed (II).

Poncirus trifoliata* **(Linnaeus) Rafinesque. Mock Orange, Trifoliate Orange, Hardy Orange. Woodland borders, fence lines, hedge rows, rich woods, escaped from cultivation. Sporadically naturalized across the panhandle and northern peninsula, Escambia to Duval counties, south to Hernando County. China.

Ptelea trifoliata **Linnaeus.** Wafer Ash, Hoptree, Common Hoptree, Stinking Ash, Skunk Bush. Bluffs, rich woodlands, forested slopes, often in association with limestone. Panhandle and northern peninsula, south to Polk County.

Severinia buxifolia* **(Poiret) Tenore. Boxthorn, Chinese Boxorange. Disturbed sites, maritime hammocks, escaped from cultivation. Sporadically naturalized, Alachua to Lee and Broward counties. Taiwan, southern China.

Triphasia trifolia* **(Burman f.) P. Wilson. Limeberry. Disturbed sites, maritime and rockland hammocks, escaped from cultivation. Southern peninsula, Broward County to the Keys. Southeast Asia, Malaysia.

!*Zanthoxylum americanum* **Miller.** Prickly Ash, Common Prickly Ash, Toothache Tree. Margins of glades, hammocks, often in association with limestone. Gadsden, Jackson, and Levy counties. State endangered.

Zanthoxylum clava-herculis **Linnaeus.** Hercules-club, Prickly Ash. Hammocks, wet woodlands, dunes, shell middens, often in association with limestone. Nearly throughout, Escambia to Miami-Dade counties; absent from the Keys.

!*Zanthoxylum coriaceum* **A. Richard.** Biscayne Prickly Ash, Leathery Prickly Ash. Coastal hammocks. Palm Beach to Miami-Dade counties. State endangered.

Zanthoxylum fagara **(Linnaeus) Sargent.** Wild Lime, Lime Prickly Ash. Hammocks. Southern peninsula, Citrus, Marion, and Volusia counties, south to the Keys.

!*Zanthoxylum flavum* **Vahl.** West Indian Satinwood, Yellowwood, Yellow Heart. Subtropical hammocks. Keys. State endangered.

Salicaceae (Willow Family)

1. Flowers borne in catkins; sepals absent
  2. Mature leaves less than 2 times longer than broad; blade broadly ovate or triangular
    3. Lower surface of the leaf whitish-hairy.....................................*Populus alba*
    3. Lower surface of the leaf not whitish-hairy
      4. Leaf blade more or less triangular, the base truncate or straight, the apex acuminate and distinctly pointed; petiole of mature leaves flattened; blade and petiole glabrous or nearly so.....................................*Populus deltoides*
      4. Leaf blade more or less ovate, the base rounded and sometimes cordate, the apex rounded or bluntly pointed; petiole round or nearly so in cross section; petiole and blade hairy when new, hairs usually retained in patches in the vein axils on the lower surface of the leaves.....................*Populus heterophylla*
  2. Mature leaves usually well over 2 times longer than broad; blade lanceolate, oblong, or narrowly oval
    5. Leaf margins mostly entire or undulate; lower surface of blade persistently and densely gray-tomentose (sometimes appearing silvery).................*Salix humulis*
    5. Leaf margins finely or sharply toothed
      6. Leaf blade distinctly grayish or glaucous beneath
        7. Larger mature leaves 3–5 cm broad, more or less long-elliptic in outline.......
        ....*Salix floridana*
        7. Larger mature leaves usually less than 3 cm broad (a few leaves on *S. caroliniana* may be to about 3.5 cm broad), more or less lanceolate or narrowly lanceolate in outline
          8. Branches erect or spreading
            9. Base of leaf blade minutely heart-shaped (most easily seen with 10× magnification); most or all leaves with conspicuous stipules..............
            ....*Salix eriocephala*
            9. Base of leaf blade not heart-shaped; stipules, if present, irregularly distributed, not subtending all or most leaves.............*Salix caroliniana*
          8. Branches pendent...*Salix babylonica*
      6. Leaf blade more or less green or yellowish green beneath...............*Salix nigra*
1. Flowers not in catkins; sepals present
  10. Fruit a capsule, flowers perfect, surface of leaves gland dotted, leaf margins finely and evenly toothed but never spiny, stipules present.........................*Casearia nitida*
  10. Fruit a berry or drupe, flowers unisexual, leaves lacking glandular dots, leaf margins entire or coarsely toothed, the marginal teeth and/or leaf tip often spiny, stipules absent
    11. Fruit < 1 cm long, leaves mostly entire, at least a few leaves with sharp, triangular marginal teeth, leaf tip sharp pointed.....................*Xylosma buxifolia*
    11. Fruit > 1 cm long, leaves crenate, the tip blunt.........................*Flacourtia indica*

Casearia nitida **(Linnaeus) Jacquin.** Smooth Casearia. Disturbed sites, southern peninsula, rarely naturalized, essentially Miami-Dade County. West Indies.

◆Flacourtia indica **(Burman f.) Merrill.** Governor's Plum, Indian Plum, Madagascar Plum, Batako Plum. Naturalized from Lee and Broward counties southward, including the Keys. Asia. FLEPPC listed (II).

Populus alba **Linnaeus.** White Poplar, Silver Poplar. Disturbed sites, escaped from cultivation. Escambia County. Eurasia.

Populus deltoides **W. Bartram ex Marshall.** Eastern Cottonwood. Floodplains and swamp margins; sometimes cultivated. Panhandle, northwestern peninsula, Escambia County, south to about Hernando County.

Populus heterophylla **Linnaeus.** Swamp Cottonwood. Floodplains. Central panhandle, Walton to Leon counties.

Salix babylonica **Linnaeus.** Weeping Willow. Disturbed sites, commonly planted, sparingly escaped from cultivation in western panhandle (Escambia County), perhaps elsewhere. Asia.

Salix caroliniana **Michaux.** Carolina Willow, Coastal Plain Willow. Moist to wet ditches, swamp and pond margins, marshes, river banks. Statewide, except the Keys.

!*Salix eriocephala* **Michaux.** Heart-leaved Willow, Missouri Willow. Mesic slopes, ditches, wet margins. Northern central panhandle, Jackson to Leon counties. State endangered.

!*Salix floridana* **Chapman.** Florida Willow. Margins of spring runs, swamps and swamp margins. Central panhandle, Jackson County, south to Lake and Orange counties. State endangered.

Salix humilis **Marshall.** Dwarf Willow, Prairie Willow. Wet margins and flatwoods, seepages in upland pinelands, dry or mesic hammocks. Eastern panhandle and north-central peninsula, from about Leon County to Alachua and Levy counties.

Salix nigra **Marshall.** Black Willow. Swamps, floodplains, pond margins. Panhandle and northern peninsula, Escambia County east to Duval and Putnam counties.

Xylosma buxifolia **A. Gray.** Mucha-gente. Everglades. Miami-Dade County.

Santalaceae (Sandalwood Family)

Santalum album **Linnaeus.** Sandalwood. Disturbed sites, rarely naturalized. Miami-Dade County. India, tropical Asia, Pacific Islands, Australia.

Sapindaceae (Soapberry Fmily)

1. Leaves simple
  2. Leaves not lobed
    3. Fruit predominantly 2-winged; leaves usually exceeding 5 cm long, with primary veins conspicuous...*Dodonaea viscosa*
    3. Fruit predominantly 3-winged; leaves usually not exceeding 5 cm long, with primary veins inconspicuous..*Dodonaea elaeagnoides*
  2. Leaves lobed and sometimes toothed
    4. Leaf margin evidently toothed, the teeth much smaller than the leaf lobes
      5. Terminal leaf lobe about equal in width from base to apex, usually shorter than half the length of the blade...*Acer rubrum*
      5. Terminal leaf lobe conspicuously narrower at base than apex, usually longer than half the length of the blade.................................*Acer saccharinum*
    4. Leaf margin not toothed, the leaf lobes sometimes appearing as large teeth
      6. Lower surface of mature leaf whitish...............................*Acer floridanum*
      6. Lower surface of mature leaf green...............................*Acer leucoderme*
1. Leaves compound
  7. Leaves palmately compound, all leaflets arising from a common point................. ....*Aesculus pavia*
  7. Leaves pinnately or bipinnately compound, or trifoliolate, leaflets not arising from a common point
    8. Leaves bipinnate.............................*Koelreuteria elegans* subsp. *formosana*
    8. Leaves pinnate or trifolioate
      9. Leaves predominantly trifoliolate, many or most leaves with only 3 leaflets
        10. Margins of leaflets usually coarsely toothed; fruit a paired samara........... ....*Acer negundo*
        10. Margins of leaflets entire; fruit round, black, fleshy........*Hypelate trifoliata*
      9. Leaves pinnate, most or all leaves with more than 3 leaflets
        11. Leaves predominately with 1–2 pairs of leaflets
          12. Flowers 4-parted; leaf axis usually winged, especially on young leaves.. ....*Melicoccus bijugatus*
          12. Flowers 5-parted; leaf axis not winged.................*Exothea paniculata*
        11. Leaves with 3 or more pairs of leaflets
          13. Leaflets coarsely and bluntly toothed.......................*Cupania glabra*
          13. Leaflets entire (sometimes wavy, but not toothed)
            14. Rachis winged......................................*Sapindus saponaria*
            14. Rachis not winged
              15. Fruit reddish orange............................*Harpullia arborea*
              15. Fruit yellowish, yellowish green, yellowish orange, or brownish
                16. Leaflets oblong, obovate, or oval, usually widest at or above the middle, apex rounded or notched........................... ....*Cupaniopsis anacardioides*

16. Leaflets lanceolate or lance-ovate, usually widest near or
below the middle, apex acuminate
17. Plant evergreen, of southern Florida.........................
....*Dimocarpus longan*
17. Plant deciduous, of northern Florida........................
....*Sapindus marginatus*

Acer floridanum (**Chapman**) **Pax.** Florida Maple. Bluffs, ravines, moist upland woods. Panhandle and northern peninsula, Okaloosa to Columbia counties, sparingly south to Polk County.

Acer leucoderme **Small.** Chalk Maple. Bluffs, ravines, upland woods. Central panhandle; Jackson, Gadsden, and Liberty counties.

Acer negundo **Linnaeus.** Box Elder, Ash-leaved Maple. Floodplains, wooded slopes, stream banks. Central panhandle and peninsula, from about the Apalachicola River south to Hillsborough, Osceola, and Brevard counties.

Acer rubrum **Linnaeus.** Red Maple, Scarlet Maple. Swamps, wet woods, moist uplands. Essentially throughout, panhandle and peninsula, south to about the Tamiami Trail in Monroe County.

Acer saccharinum **Linnaeus.** Silver Maple. Floodplains, bottomlands, riverbanks; widely planted. Escambia County (perhaps naturalized here), central panhandle (Jackson and Liberty counties to Leon County), Citrus County.

Aesculus pavia **Linnaeus.** Red Buckeye. Slopes, bottoms, ravines, bluffs, hammocks, rich mesic woods. Panhandle and northern peninsula, Escambia to Nassau counties, south to Sumter, Lake, and Orange counties.

!*Cupania glabra* **Swartz.** Florida Cupania, American Toadwood. Subtropical hammocks. Keys. State endangered.

*◆*Cupaniopsis anacardioides* (**A. Richard**) **Radlkofer.** Carrotwood. Disturbed sites, scrubs, sandhill, hammocks, flatwoods, dunes, marshes, mangrove-dominated wetlands, cypress swamps. Southern peninsula, Pinellas and Volusia counties southward; absent from the Keys. Australia. FLEPPC listed (I).

Dimocarpus longan **Loureiro.** Longan. Disturbed sites, escaped from cultivation. Palm Beach County. China.

!*Dodonaea elaeagnoides* **Rudolph ex Ledebour & Alderstam.** Smallfruit Varnish Leaf, Keys Hopbush. Coastal hammocks and woodlands. Keys. State endangered.

Dodonaea viscosa (**Linnaeus**) **Jacquin.** Varnish Leaf, Florida Hop Bush. Coastally along the central peninsula, more widespread in the southern peninsula, St. Johns and Hernando counties southward, including the Keys.

Exothea paniculata (**Jussieu**) **Radlkofer ex Durand.** Inkwood, Butterbough. Hammocks, shell mounds. Volusia County southward along the east coast, Collier County southward along the west coast; generally throughout the southern peninsula, including the Keys.

Harpullia arborea (**Blanco**) **Radlkofer.** Tulipwood, Uas. Disturbed sites, hammocks, escaped from cultivation. Miami-Dade County. Asia, Australia, Philippines.

!*Hypelate trifoliata* **Swartz.** White Ironwood. Hammocks. Miami-Dade County and the Keys. State endangered.

*◆*Koelreuteria elegans* (Seemann) A. C. Smith subsp. *formosana* (Hayata) F. G. Meyer. Golden Rain Tree, Flamegold. Disturbed sites, roadsides, escaped from cultivation. Sporadically naturalized from about Putnam County southward to Broward County. Taiwan. FLEPPC listed (II).

Melicoccus bijugatus Jacquin. Spanish Lime. Disturbed sites, escaped from cultivation. Potentially naturalized from Palm Beach County southward, including the Keys. Central and northern South America.

Sapindus marginatus Willdenow. Florida Soapberry. Coastal woodlands, hammocks, shell middens, often in association with limestone. Panhandle, south to about Lee County.

Sapindus saponaria Linnaeus. Tropical Soapberry, Soapberry, Wingleaf Soapberry. Hammocks, coastal scrub. Sumter County southward, including the Keys.

Sapotaceae (Sapodilla Family)

1. Plant thorny, either with axillary thorns or thorn-tipped branches (thorn-tipped short shoots may lose the thorny tip with age)
    2. Lower surface of mature leaves glabrous or nearly so (except sometimes along the midvein)
        3. Leaf apex acute or acuminate...................................*Sideroxylon lycioides*
        3. Leaf apex rounded or blunt
            4. Veins on lower surface of leaf blade neither prominent nor conspicuous..........
            ....*Sideroxylon celastrinum*
            4. Veins on lower surface of leaf blade prominent and conspicuous
                5. Petioles at least sparsely hairy, usually densely so...............................
                ....*Sideroxylon lanuginosum*
                5. Petioles glabrous or nearly so
                    6. Fruit 9–16 mm in diameter...........................*Sideroxylon lycioides*
                    6. Fruit 4–9 mm in diameter.........*Sideroxylon reclinatum* subsp. *reclinatum*
    2. Lower surface of leaf hairy
        7. Hairs on lower leaf surface tightly appressed to blade, densely matted, usually silvery or silvery gray...............................*Sideroxylon alachuense*
        7. Hairs on the lower leaf surface not with the above combination
            8. Stem of the current season glabrous
                9. Many or most leaves longer than 8 cm...................*Sideroxylon lycioides*
                9. Few or no leaves exceeding about 7 cm long
                    10. Veins of the upper leaf surface faint, not at all raised, often impressed, those forming the islets pale but not significantly contrasting in color with the tissue enclosed within the islets (requires at least 10× magnification)...*Sideroxylon thornei*
                    10. Veins of the upper leaf surface usually slightly raised, somewhat bony in

color and contrasting with the tissue enclosed within the islets..............
......................................*Sideroxylon reclinatum* subsp. *austroflloridense*
8. Stem of the current season hairy
    11. Lower surface of the leaf with densely matted with tightly appressed hairs, these lustrous brown and obscuring the veins.................*Sideroxylon tenax*
    11. Hairs on lower surface of leaf not as above
        12. Mature plant usually less than 1 m tall, forming extensive clonal colonies; lower surface of mature leaves usually glabrous (very hairy when new)..*Sideroxylon rufohirtum*
        12. Mature plants well over 1 cm tall, not forming clonal colonies; lower surface of mature leaves densely hairy, many or most often feltlike to the touch..*Sideroxylon lanuginosum*
1. Plant not thorny
    13. Lower surface of many or most leaves coppery brown with densely matted hairs..*Chrysophyllum oliviforme*
    13. Lower surface of leaves not copper brown, glabrous or nearly so
        14. Flowers with 6 or 8 sepals in 2 whorls
            15. Sepals 8, in 2 whorls of 4
                16. Leaf apex notched; petiole 1–1.5 cm long; pedicel 4–5 cm long; fruit 3–5 cm in diameter...................................*Mimusops coriacea*
                16. Leaf apex not notched; petiole 1.5–2.5 cm long; pedicel to about 1 cm long; fruit 1.5–2.5 cm in diameter.....................*Mimusops elengi*
            15. Sepals 6, in 2 whorls of 3
                17. Many or most leaves notched at the apex; fruit to about 4 cm in diameter..........................*Manilkara jaimiqui* subsp. *emarginata*
                17. Most leaves rounded or acuminate at apex, rarely notched; fruit 5–10 cm in diameter..*Manilkara zapota*
        14. Flowers with 4–6 sepals in a single whorl
            18. Young stems and pedicels glabrous; sepals 4.5–11 mm long; fruit 5–7 cm long..*Pouteria campechiana*
            18. Young stems and pedicels densely hairy; sepals not exceeding about 2 mm long; fruit 2–2.5 cm long
                19. Petioles 14–51 mm long; berries yellow or orange at maturity..............
....*Sideroxylon foetidissimum*
                19. Petioles 1–14 mm long; berries black at maturity..........................
....*Sideroxylon salicifolium*

!*Chrysophyllum oliviforme* **Linnaeus.** Satinleaf. Subtropical hammocks. Brevard County, south along the coast, generally throughout the southern peninsula from Hendry County south, including the Keys. State threatened.
!*Manilkara jaimiqui* **(C.Wright ex Grisebach) Dubard** subsp. *emarginata* **(Linnaeus) Cronquist.** Wild Dilly. Hammocks. Collier and Miami-Dade counties southward, including the Keys. State threatened.
*◆*Manilkara zapota* **(Linnaeus) P. Royen.** Sapodilla. Disturbed sites, hammocks,

escaped from cultivation. Southern peninsula, Lee and Palm Beach counties southward. Mexico, Central America. FLEPPC listed (I).

*Mimusops coriacea (**A. de Candolle**) **Miquel.** Monkey's Apple. Disturbed sites, escaped from cultivation. Broward and Miami-Dade counties. Mascarenes.

*Mimusops elengi **Linnaeus.** Spanish Cherry, Kabiki, Bakul. Disturbed sites, escaped from cultivation. Broward County. India, Sri Lanka, Malaysia.

*Pouteria campechiana (**Kunth**) **Baehni.** Egg Fruit, Canistel. Disturbed sites, hammocks, escaped from cultivation. Miami-Dade County and the Keys. Mexico, Central America, West Indies.

•!*Sideroxylon alachuense* **L. C. Anderson.** Silver Buckthorn, Silver Bully, Clark's Buckthorn. Hammocks. Nassau, Alachua, Marion, Lake, and Orange counties. State endangered.

Sideroxylon celastrinum (**Kunth**) **T. D. Pennington.** Saffron Plum. Hammocks, salt flats. Largely coastal from Levy and Brevard counties southward, including the Keys.

Sideroxylon foetidissimum **Jacquin.** False Mastic. Coastal Hammocks. Coastal counties, Manatee and Volusia counties southward, including the Keys.

Sideroxylon lanuginosum **Michaux.** Gum Bully. Dry upland woods, moist and dry hammocks. Panhandle and northern peninsula, Escambia to St. Johns counties, south to Pinellas, Hillsborough, Lake, and Orange counties.

!*Sideroxylon lycioides* **Linnaeus.** Buckthorn Bully, Gopherwood Buckthorn. Floodplains, bottomlands, moist calcareous woods. Sporadic across the panhandle and northern peninsula, Escambia to Clay and south to Lake and Orange counties. State endangered.

Sideroxylon reclinatum **Michaux subsp.** *reclinatum.* Florida Bully. Bluffs, ravines, riverbanks, calcareous and mesic hammocks, floodplains. Essentially or potentially throughout the state, including the Keys.

•*Sideroxylon reclinatum* **Michx. subsp.** *austrofloridense* (**Whetstone**) **Kartesz & Gandhi.** Florida Bully. Calcareous glades. Miami-Dade County.

•*Sideroxylon rufohirtum* **Herring & Judd.** Rufous Florida Bully. Hammocks. North- and west-central peninsula, Suwannee and Columbia counties south to Manatee and Orange counties.

Sideroxylon salicifolium (**Linnaeus**) **Lamarck.** Willow Bustic, Bustic, White Bully. Hammocks, margins of pinelands. South Florida, Martin and Collier counties southward, including the Keys.

Sideroxylon tenax **Linnaeus.** Tough Bully. Coastal dunes, interior scrub. Nassau and Levy counties south to Collier and Miami-Dade counties.

!*Sideroxylon thornei* (**Cronquist**) **T. D. Pennington.** Georgia Bully, Thorne's Buckthorn. Wooded drainages, wet woodland depressions, stream swamps, typically where water stands much of the time. Panhandle, Escambia, Santa Rosa, Holmes, Jackson, Gulf, and Franklin counties. State endangered.

Schisandraceae (Starvine Family)

!*Schisandra glabra* (**Brickell**) **Rehder.** Star-vine, Bay Star Vine, Wild Sarsaparilla, Schisandra. Rich woods, bluffs, moist slopes. Central panhandle, Holmes and Jackson to Leon and Wakulla counties. State endangered.

Schoepfiaceae (Schoepfia Family)

Schoepfia chrysophylloides (**A. Richard**) **Planchon.** Gulf Graytwig, Graytwig, White-wood. Hammocks. Volusia County southward along the east coast to the Keys; more generally distributed across the southern peninsula, Hendry and Collier counties.

Simaroubaceae (Quassia Family)

1. Leaves with an odd number of leaflets (terminal leaflet present); leaflets toothed at base, with a flattened gland on the lower surface near the tip of one or more of the teeth..*Ailanthus altissima*
1. Leaves with an even number of leaflets (true terminal leaflet absent); Leaflets not toothed at base..*Simarouba glauca*

*Ailanthus altissima (**Miller**) **Swingle.** Tree-of-Heaven. Disturbed sites, roadsides, fence lines. Sporadically naturalized, Jackson, Franklin, Jefferson, Alachua, and Hillsborough counties. Asia.
Simarouba glauca **de Candolle.** Paradise Tree, Bitterwood. Coastal and inland hammocks. Brevard and Collier counties southward including the Keys.

Smilacaceae (Greenbriar or Smilax Family)

1. Leaves conspicuously whitish beneath...*Smilax glauca*
1. Leaves not conspicuously whitish beneath
   2. Plant trailing; stem lacking prickles or spines; stems and petioles shaggy hairy; lower surface of the leaf densely hairy.............................*Smilax pumila*
   2. Plant lacking the above combination of characters
      3. Margins and midvein on lower leaf surface often bearing prickles
         4. Prickles and lower portion of main stem roughened with tawny or orangish scales (scurfy); the base of many leaves lobed....................*Smilax bona-nox*
         4. Prickles glabrous; base of leaf not lobed........................*Smilax havanensis*
      3. Prickles and lower portion of main stem lacking scales; leaf margins and midvein usually lacking prickles
         5. Midvein on lower surface of leaf distinctly raised and more pronounced than the lateral veins, the lateral veins usually obscure..................*Smilax laurifolia*
         5. Midvein on lower surface of leaf little if any more pronounced than the lateral veins, the lateral veins clearly evident
            6. Prickles on main shoots very slender, needlelike, usually lustrous dark brown or black...*Smilax tamnoides*
            6. Prickles neither needlelike nor lustrous brown or black
               7. Leaves lanceolate..*Smilax smallii*
               7. Leaves oblong, ovate, hastate, or nearly linear
                  8. Many leaves bearing two marginal veins, these often closely set and appearing as a conspicuous band with a central groove; petiole usually green...*Smilax auriculata*
                  8. Leaves lacking a thickened band at the margin; petioles usually reddish
                     9. Plant deciduous, usually of swamps and other wetlands, often climbing over standing water; berries red..............*Smilax walteri*
                     9. Plant evergreen, usually growing in well-drained locations or where only briefly flooded; berries blue-black............*Smilax rotundifolia*

Smilax auriculata **Walter.** Earleaf Greenbrier, Greenbrier, Catbrier. Sandhills, scrub oak ridges, sandy sites along the coast, flatwoods, wet or moist woodlands, hammocks. Essentially throughout, except the Keys.

Smilax bona-nox **Linnaeus.** Saw Greenbrier, Catbrier, Greenbrier. Upland and lowland woods, wet pinelands, hammocks, disturbed sites, fields. Essentially throughout, including the Keys.

Smilax glauca **Walter.** Wild Sarsaparilla, Cat Greenbrier. Upland and lowland woods, flatwoods, hammocks, disturbed sites, fields. Panhandle, northern peninsula, south to Sarasota, Polk, and Brevard counties.

!*Smilax havanensis* **Jacquin.** Everglades Greenbrier. Hammocks, pinelands. Southernmost peninsula, Collier and Miami-Dade counties and the Keys. State threatened.

Smilax laurifolia **Linnaeus.** Bamboo Vine, Blaspheme Vine, Laurel Greenbrier. Wet and dry pinelands, bay swamps, stream banks, usually in wet sites. Essentially throughout, except the Keys.

***Smilax pumila* Walter.** Dwarf Smilax, Sarsaparilla Vine. Moist or dry woods, sand pine—oak scrub, sandy coastal woods. Panhandle and northern peninsula, south to Sarasota, DeSoto, Highlands, Osceola, and Brevard counties.

***Smilax rotundifolia* Linnaeus.** Bullbrier, Horsebrier, Common Greenbrier, Roundleaf Greenbrier. Moist and dry woods, river banks, pond and lake margins. Sporadic in the panhandle and northern peninsula, Escambia to Nassau counties, south to Alachua and Volusia counties.

***Smilax smallii* Morong.** Jackson Vine, Jackson Brier, Lanceleaf Greenbrier. Rich woods, hammocks, well-drained woodlands. Panhandle, sparingly south to Hillsborough and Highlands counties.

***Smilax tamnoides* Linnaeus.** Bristly Greenbrier, Hogbrier. Uplands, old fields, clearings, stream banks, moist hammocks. Panhandle and peninsula, south to Collier and Palm Beach counties.

***Smilax walteri* Pursh.** Coral Greenbrier. Marshes, cypress swamps, pineland depressions, floodplains, often over or near standing water. Panhandle and northern peninsula, south to Sarasota, Glades, Osceola, and Brevard counties.

Solanaceae (Nightshade Family)

1. Plant a vine or vinelike
   2. Flowers large, cup shaped, usually exceeding 16 cm long...............*Solandra nitida*
   2. Flowers smaller, usually less than 16 cm long...................*Solanum seaforthianum*
1. Plant a tree or shrub, sometimes scrambling with vinelike branches
   3. Flowers at least 16 cm long.......................................*Brugmansia suaveolens*
   3. Flowers smaller, usually less than 16 cm long
      4. Plant armed with prickles
         5. Leaves sessile...*Solanum jamaicense*
         5. Leaves distinctly stalked
            6. Plant a sprawling, straggly, vinelike shrub; corolla deeply lobed, petals fused only at the base...*Solanum tampicense*
            6. Plant a shrub or small tree; petals fused above the base
               7. Plant hairy with stellate pubescence
                  8. Mature fruit usually red, not exceeding about 1 cm in diameter...........
                  ....*Solanum bahamense*
                  8. Mature fruit usually yellow, greater than 1 cm in diameter................
                  ....*Solanum torvum*
               7. Plant essentially glabrous, hairs, if present, not stellate.....................
               ....*Solanum diphyllum*
      4. Plant unarmed, or essentially so
         9. Corolla mostly rotate, the petals more or less radiating from the center of the

flower like the spokes of a wheel

    10. Flowers borne 1–3 at or near the leaf nodes or in axillary fascicles

        11. Anthers not united, distinctly separate; fruit longer than broad, distinctly pepper-like

            12. Fruit less than 1.5 cm long, usually 1 per node.....*Capsicum annuum*

            12. Fruit usually greater than 1.5 cm long, usually 2 or more per node...
....*Capsicum frutescens*

        11. Anthers fused or appearing so; fruit round, tomato-like......................
....*Solanum pseudocapsicum*

    10. Flowers borne in several- or many-flowered inflorescences

        13. Fruit hairy with stellate pubescence

            14. Flowers purple or lilac; leaves on young branches subtended in the axil with 1–2 conspicuous but smaller leaves..*Solanum mauritianum*

            14. Flowers white; leaves not subtended by smaller leaves.................
....*Solanum erianthum*

        13. Fruit glabrous

            15. Tissue of leaf blade extending onto the petiole; petiole slightly winged for most of its length.....................*Solanum umbellatum*

            15. Tissue not extending onto the petiole; petiole not winged.............
....*Solanum donianum*

  9. Corolla tubular, campanulate, salverform, or funnelform

    16. Leaves linear, more or less succulent....................*Lycium carolinianum*

    16. Leaves elliptic or lanceolate, not succulent

        17. Fruit a capsule...*Nicotiana glauca*

        17. Fruit a berry

            18. Flowers white

                19. Corolla lobes reflexed; fruit black.............*Cestrum dirunum*

                19. Corolla lobes erect or spreading; fruit white........................
....*Cestrum nocturnum*

            18. Flowers bright yellow or creamy yellow..............*Cestrum parqui*

***Brugmansia suaveolens* (Humboldt & Bonpland ex Willdenow) Berchtold & J. Presl.** Angel's Trumpet. Disturbed sites, widely planted, escaped from cultivation. Rarely naturalized in the central and southern peninsula (Hernando and Broward counties). South America.

***Capsicum annuum* Linnaeus var. *glabriusculum* (Dunal) Heiser & Pickersgill.** Bird Pepper, Red Pepper, Cayenne Pepper. Hammocks, shell middens, disturbed sites. Sporadic in the peninsula, Levy, Alachua, and Clay counties south to the Keys; more widespread in the southernmost peninsula.

***Capsicum frutescens* Linnaeus.** Tabasco Pepper. Disturbed sites. Central and southern peninsula, potential from about Citrus, Orange, and Volusia counties southward, including the Keys.

***◆Cestrum diurnum* Linnaeus.** Day Jessamine, Dayflowering Jessamine. Disturbed sites, escaped from cultivation. Southern peninsula, potentially Hillsborough and Palm

Beach counties southward, including the Keys. Tropical America, West Indies. FLEPPC listed (II).

*Cestrum nocturnum **Linnaeus.** Nightflowering Jessamine. Disturbed sites, escaped from cultivation. Southwestern peninsula, Pinellas and Hillsborough counties sporadically south to the Keys.

*Cestrum parqui **L'Héritier.** Chilean Jessamine. Disturbed sites, escaped from cultivation. Hernando County. South America.

Lycium carolinianum **Walter.** Christmas Berry, Matrimony Vine, Carolina Desert-thorn. Beaches, shell middens, sandy shores, usually nearly salt- or brackish water. Coastally, Gulf and Duval counties southward to the Keys.

*Nicotiana glauca **Graham.** Tree Tobacco. Disturbed sites, escaped from cultivation. Sparingly naturalized, Escambia and Marion counties. South America.

*Solandra nitida **Zuccagni.** Chalicevine. Disturbed sites, escaped from cultivation. Miami-Dade County. Mexico.

Solanum bahamense **Linnaeus.** Bahama Nightshade, Cankerberry. Hammocks, dunes. Southeastern peninsula, Martin County south to the Keys. Includes *Solanum bahamense* Linnaeus var. *rugelii* D'Arcy.

*◆Solanum diphyllum **Linnaeus.** Twoleaf Nightshade. Disturbed sites, escaped from cultivation. Peninsula, Alachua County, increasingly widespread southward to Lee and Miami-Dade counties. Mexico. FLEPPC listed (II).

!Solanum donianum **Walpers.** Mullein Nightshade. Hammocks. Southernmost peninsula, Collier and Miami-Dade counties southward to the Keys. State threatened.

Solanum erianthum **D. Don.** Potato Tree, Mullein Nightshade. Thickets, disturbed sites, hammock margins. St. Johns, Putnam, and Hernando counties southward to the Keys; more common in the southernmost peninsula.

*◆Solanum jamaicense **Miller.** Jamaican Nightshade. Disturbed sites. Central peninsula, Seminole, Polk, Osceola, Highlands, and St. Lucie counties. West Indies. FLEPPC listed (II).

*Solanum mauritianum **Scopoli.** Earleaf Nightshade. Disturbed woodlands. Pasco County.

*Solanum pseudocapsicum **Linnaeus.** Jerusalem Cherry. Disturbed sites, escaped from cultivation. Sparsely naturalized, Leon and Jefferson counties.

*Solanum seaforthianum **Andrews.** Brazilian Nightshade. Disturbed sites, escaped from cultivation. Central and southern peninsula, Citrus and Brevard counties south to Collier and Miami-Dade counties.

*◆Solanum tampicense **Dunal.** Aquatic Soda Apple. Floodplains. Southwestern peninsula (Highlands, DeSoto, Charlotte, Lee, and Glades counties) and the Keys. West Indies, Mexico, Central America. FLEPPC listed (I).

*◆Solanum torvum **Swartz.** Turkeyberry. Disturbed sites, fields, roadsides. Potentially throughout the peninsula, especially from Lee, Glades, and Palm Beach counties southward. West Indies. FLEPPC listed (II).

*Solanum umbellatum **Miller.** Disturbed sites. Miami-Dade County. Mexico.

Staphyleaceae (Bladdernut Family)

!*Staphylea trifolia* Linnaeus. Bladdernut, American Bladdernut. Floodplains, rich wooded slopes. Gadsden and Liberty counties. State endangered.

Strychnaceae (Strychnos Family)

**Strychnos spinosa* Lamarck. Wood Orange, Natal Orange. Disturbed sites. Sporadically and sparsely naturalized in the southern peninsula, Hillsborough, Highlands, and Miami-Dade counties.

Styracaceae (Storax Family)

1. Corolla 4-lobed; fruit elongate, winged
    2. Blades of mature leaves long-elliptic, usually widest near the middle; flower petals fused for at least half their length; fruit with 4 more or less equal wings................
    ....*Halesia carolina*
    2. Blades of mature leaves broadly oval, often widest above the middle; flower petals free nearly to base; fruit with 2 large wings and 2 small wings, usually appearing as 2-winged
        3. Flower petals 1–1.5 cm long...*Halesia diptera* var. *diptera*
        3. Flower petals 2–3 cm long.....................................*Halesia diptera* var. *magniflora*
1. Corolla 5-lobed or more; fruit rounded, not winged
    4. Mature leaves not exceeding about 8 cm long and 4 cm broad; flower petals strongly recurved..*Styrax americana*
    4. Mature leaves or many of them greater than 8 cm long and greater than 4 cm broad; petals not reflexed..*Styrax grandifolius*

Halesia carolina Linnaeus. Little Silverbell, Carolina Silverbell. Sandy woods and slopes, bluff ridges, hammocks, floodplains. Panhandle east to Columbia, Gilchrist, Levy, and Citrus counties.
Halesia diptera Ellis var. *diptera*. Two-winged Silverbell. Rich woods and slopes, hammocks, ravines, bluffs, floodplain woodlands. Panhandle, Escambia to Jefferson counties, disjunct to Suwannee County.
Halesia diptera Ellis var. *magniflora* Godfrey. Large-flowered Two-winged Silverbell. Rich woods, hammocks, slopes, ravines, floodplains. Panhandle, Escambia County to about Leon County.

Styrax americanus Lamarck. American Snowbell. Calcareous hammocks, rich woods, swamps, often in wet habitats, often with standing water. Panhandle and northern peninsula, south to Charlotte and Osceola counties.

Styrax grandifolius **Aiton.** Bigleaf Snowbell. Rich bluffs and woodlands, ravines, calcareous hammocks, well-drained woods. Panhandle, Santa Rosa to Madison and Taylor counties, disjunct to Duval County.

Surianaceae (Bay Cedar Family)

Suriana maritima **Linnaeus.** Bay Cedar. Beaches, dunes. Coastal strand, southern peninsula, Pinellas, Hillsborough, and Indian River counties, south to the Keys.

Symplocaceae (Sweetleaf Family)

Symplocos tinctoria **(Linnaeus) L'Héritier de Brutelle.** Sweetleaf, Common Sweetleaf, Horse Sugar. Sandhills, ravines, rich slopes, moist hammocks, floodplains, bottomlands, margins of sinkholes. Panhandle and northern peninsula, south to about Levy and Alachua counties; disjunct to Hillsborough County.

Tamaricaceae (Tamarisk Family)

*Tamarix canariensis **Willdenow.** Canary Island Tamarisk. Beaches, dry coastal woods. Sporadic in Florida, Franklin and Duval counties. Canary Islands.

Taxaceae (Yew Family)

1. Leaves stiff, tip sharp and piercing to the touch; branches whorled, produced in layers along the trunk...*Torreya taxifolia*
1. Leaves flexible, tip not piercing to the touch; branches not whorled.......*Taxus floridana*

•!*Taxus floridana* **Nuttall ex Chapman.** Florida Yew. Drier portions of slopes and ravines. Eastern edge of the Apalachicola River, Gadsden and Liberty counties. State endangered.
!*Torreya taxifolia* **Arnott.** Florida Torreya, Torreya, Stinking Cedar, Gopherwood. Bluffs and ravines, in association with the Apalachicola River, central panhandle. Liberty, Gadsden, and Jackson (where historical) counties. State and federally endangered.

Theaceae (Tea Family)

1. Plant evergreen, leaves stiff, leathery, the margins coarsely and bluntly toothed.........
....*Gordonia lasianthus*
1. Plant deciduous, leaves thin, pliable, the margins finely and obscurely toothed and often ciliate...*Stewartia malacodendron*

Gordonia lasianthus **(Linnaeus) Ellis.** Loblolly Bay. Swamps, bogs, margins of wet flatwoods. Panhandle, northern and south-central peninsula, Escambia and Nassau counties south to Sarasota, Glades, and Palm Beach counties.
!*Stewartia malacodendron* **Linnaeus.** Stewartia, Silky Camellia. Moist wooded slopes, rich bluffs, creek banks. Panhandle, Escambia to Leon counties. State endangered.

Theophrastaceae (Theophrasta Family)

1. Leaf tip blunt or acute and often spine-tipped........................*Bonellia macrocarpa*
1. Leaf tip rounded or notched, not spine-tipped (sometimes with a tiny toothlike extention)
    2. Leaf blade not exceeding 3 cm long and 1 cm broad.................*Jacquinia keyensis*
    2. Leaf blade or most of them exceeding 3 cm long and 1.5 cm broad.....................
    ....*Jacquinia arborea*

Bonellia macrocarpa (**Cavanilles**) **Ståhl & Källersjö.** Cudjoewood. Margins of mangrove-dominated wetlands. Miami-Dade County. Tropical America.
Jacquinia arborea **Vahl.** Braceletwood. Disturbed sites, margins of mangrove-dominated wetlands. Broward County south to the Keys. Tropical America.
!*Jacquinia keyensis* **Mez.** Joewood. Coastal hammocks. Lee and Miami-Dade counties and the Keys. State threatened.

Thymelaeaceae (Mezereum Family)

!*Dirca palustris* **Linnaeus.** Leatherwood, Eastern Leatherwood. Ravine slopes, rich woods, river banks, bluff forests. Mostly in association with the Apalachicola River, Jackson, Gadsden, and Liberty counties. State endangered.

Turneraceae (Turnera Family)

Turnera ulmifolia **Linnaeus.** Yellow Alder, Ramgoat Dashalong. Disturbed sites, escaped from cultivation. Peninsula, Volusia, and Hernando counties south to the Keys. Tropical America.

Ulmaceae (Elm Family)

1. Bark scaly and flaking, exposing reddish brown inner bark; leaves mostly widest at base, symmetrical or nearly so on either side of the midrib; fruit a soft, wingless, burrlike nut...*Planera aquatica*
1. Bark not scaly and flaky, usually furrowed and ridged, sometimes more or less smooth and mottled gray, brown, and rusty; leaves usually broadest above the base, tissue moderately to strongly asymmetrical on either side of the midrib; fruit a samara
   2. Leaf margins mostly singly toothed; bark smooth or slightly scaly, mottled...........
   ....*Ulmus parvifolia*
   2. Leaf margins doubly toothed; bark ridged, furrowed, and moderately scaly, not mottled
      3. Petiole less than 4 mm long; young branches usually bearing corky wings; apex of leaf not fitted with a short, sharp point

  4. Leaf blade scabrous and rough to the touch above, the apex rounded or blunt; samara 8–10 mm long, its surfaces sparsely hairy...............*Ulmus crassifolia*
  4. Leaf blade smooth above, the apex pointed; samara not exceeding 8 mm long, its surfaces copiously hairy..*Ulmus alata*
  3. Petiole at least 4 mm long; young branches lacking corky wings; apex of the leaf fitted with a short, sharp point
    5. Leaf blade usually strongly to moderately asymmetrical on either side of the midrib, the upper surface of mature leaves often smooth to the touch (some may be harshly scabrid); bud scales glabrous or minutely hairy; surfaces of the samara glabrous, the margins ciliate
      6. Leaf base strongly oblique, larger leaves 10–15 cm long....*Ulmus americana*
      6. Leaf base moderately oblique, larger leaves 7–10 cm long....*Ulmus floridana*
    5. Leaf blade moderately or slightly asymmetrical on either side of the midrib, the upper surface of mature leaves always finely scabrid; bud scales with reddish brown hairs; surfaces of the samara hairy over the seed-bearing portion, the margins not ciliate...*Ulmus rubra*

Planera aquatica **J. F. Gmelin.** Planertree, Water Elm. Riverbanks, backwaters, swamps, margins of oxbow lakes and wet depressions, floodplains. Panhandle and northern peninsula, Escambia to Nassau counties, south to Alachua and Levy counties.

Ulmus alata **Michaux.** Winged Elm, Cork Elm. Floodplains, slopes, and rich, well-drained forests. Panhandle and northern peninsula, Escambia to Duval counties, south to Pasco and Seminole counties.

Ulmus americana **Linnaeus.** American Elm. Floodplain forests, moist woodlands. Panhandle and peninsula, Escambia to Duval counties, south to the central peninsula.

Ulmus crassifolia **Nuttall.** Cedar Elm. Floodplains. Mostly in association with the Suwannee River, Suwannee to Marion and Hernando counties.

Ulmus floridana **Chapman.** Florida Elm. River banks, floodplains, wet hammocks. Panhandle and peninsula, south to Collier County.

*Ulmus parvifolia **Jacquin.** Chinese Elm, Drake Elm. Disturbed sites, escaped from cultivation. Sporadically naturalized, Escambia and Highlands counties and presumably elsewhere. Asia.

Ulmus rubra **Muhlenberg.** Red Elm, Slippery Elm. Rich woods, bluffs, often in association with calcareous woodlands. Panhandle, Holmes to Jefferson counties.

Verbenaceae (Vervain Family)

1. Plant a woody vine...*Petrea volubilis*
1. Plant a tree or shrub
  2. Flowers and fruit embedded in a fleshy rachis
    3. Veins on upper surface of leaf conspicuously depressed, but not quilted; plant a low herb or suffrutescent shrub; marginal teeth blunt, averaging in number about 24 per leaf..*Stachytarpheta jamaicensis*
    3. Veins on upper surface of leaf conspicuously quilted with central, lateral, and secondary veins deeply impressed; marginal teeth acute, averaging about 38 per leaf..*Stachytarpheta cayennensis*
  2. Flowers and fruit not embedded in a fleshy rachis
    4. Inflorescence a drooping, axillary raceme
      5. Leaf margins entire; most blades exceeding 5 cm long; flowers white; fruit reddish brown or blackish..*Citharexylum spinosum*
      5. Margins of at least some leaves toothed; most blades not exceeding 5 cm long; flowers blue or purple; fruit yellow...*Duranta erecta*
    4. Inflorescence a dense axillary spike or flat-topped head
      6. Inflorescence a dense axillary spike.............................*Lantana canescens*
      6. Inflorescence a dense flat-topped head
        7. Head subtended by and nestled in or above a cuplike series of bracts
          8. Corolla not exceeding 3 mm long, white or pale blue..*Lantana involucrata*
          8. Corolla 8–20 mm long, magenta or lilac.............*Lantana montevidensis*
        7. Head not subtended by a series of bracts
          9. Plant erect or sprawling, mature plants exceeding 50 cm tall; blades of most leaves more than 3 cm long; flowers usually multi-colored..*Lantana camara*
          9. Plant prostrate or decumbent, not exceeding 50 cm tall; blades of most leaves not exceeding 3 cm long; all flowers solid yellow...*Lantana depressa*

Citharexylum spinosum **Linnaeus.** Fiddlewood, Florida Fiddlewood. Hammocks, pinelands. East coast, Brevard to Monroe and Miami-Dade counties and the Keys.
*Duranta erecta **Linnaeus.** Golden Dewdrops, Pigeon Berry, Sky Flower. Disturbed sites, escaped from cultivation. Widely cultivated nearly throughout, natualized mostly in the central and southern peninsula and the Keys.West Indies.
*◆Lantana camara **Linnaeus.** Lantana, Shrub Verbena. Disturbed sites, sandhills, flatwoods, hammocks, escaped from cultivation. West Indies. FLEPPC listed (I).
!Lantana canescens **Kunth.** Hammock Shrub Verbena. Hammocks. Miami-Dade County. State endangered.
•!Lantana depressa **Small.** Rockland Shrub Verbena, Pineland Lantana. Pine rocklands.

Miami-Dade County. State endangered.

Lantana involucrata **Linnaeus.** Buttonsage. Hammocks, dunes. Coastal counties, central and southern peninsula, Pinellas, Hillsborough, and Brevard counties southward, including the Keys.

Lantana montevidensis* **(Sprengel) Briquet. Trailing Shrub Verbena. Disturbed sites, escaped from cultivation. Widely planted and sporadically naturalized, Jackson and Franklin counties south to Monroe and Miami-Dade counties, not including the Keys.

*Petrea volubilis **Linnaeus.** Queenswreath. Disturbed sites, escaped from cultivation. Scarcely naturalized, Lee County.

*◆*Stachytarpheta cayennensis* **(Richard) Vahl.** Nettleleaf Velvetberry. Disturbed sites, escaped from cultivation. Southern peninsula, Miami-Dade and Collier counties. French Guiana. FLEPPC listed (II).

Stachytarpheta jamaicensis **(Linnaeus) Vahl.** Blue Porterweed, Joee. Dunes, shell middens, pine rocklands, roadsides, disturbed sites. Central and southern peninsula, Hillsborough, Osceola, and Brevard counties southward including the Keys.

*Stachytarpheta × *intercedens* **Danser.** (*S. jamaicensis* × *S. cayennensis*).

Veronicaceae (Speedwell Family)

Capraria biflora **Linnaeus.** Goatweed. Beaches, margins of coastal hammocks. South Florida, Martin, and Lee counties southward including the Keys.

Viscaceae (Mistletoe Family)

1. Twigs round in cross section, fruit white or yellowish white.................................
....*Phoradendron leucarpum*
1. Twigs angled in cross section, fruit yellowish or orange...........*Phoradendron rubrum*

Phoradendron leucarpum **(Rafinesque) Reveal & M. C. Johnston.** Mistletoe, Oak Mistletoe. Parasitic on tree branches, upland and lowland woods. Panhandle and peninsula, south to Collier, Okeechobee, and St. Lucie counties.

!*Phoradendron rubrum* **(Linnaeus) Grisebach.** Mahogany Mistletoe. Parasitic on tree branches, subtropical hammocks. Miami-Dade County and the Keys. State endangered.

Vitaceae (Grape Family)

1. Leaves compound
    2. Leaves and branches more or less succulent or subsucculent............*Cissus trifoliata*
    2. Leaves not at all succulent
        3. Leaves palmately compound............................*Parthenocissus quinquefolia*
        3. Leaves predominately bipinnate or tripinnate....................*Ampelopsis arborea*
1. Leaves simple
    4. Leaves leathery (coriaceous)...*Cissus verticillata*
    4. Leaves thinner, not leathery
        5. Pith white; petals free, not fused at their tips and falling separately................
        ....*Ampelopsis cordata*
        5. Pith brown; petals fused at the apex, separating and falling as a single cap-like unit
            6. Tip of the tendrils simple, not branched or divided..............*Vitis rotundifolia*
            6. Tip of the tendrils forked, divided into 2–3 branches
                7. Lower surface of the leaf, fruit, and leaf nodes usually glaucous, the leaf
                glaucescence often obscured by tangled hairs....................*Vitis aestivalis*
                7. Lower surface of leaf often hairy but not glaucous; leaf nodes not glaucous;
                fruit at most only slightly glaucous
                    8. Lower surface of leaf densely and evenly white- or rusty-tomentose,
                    distinctly concealing the blade surface if not the veins...*Vitis shuttleworthii*
                    8. Lower surface of leaf blade not densely white- or rusty-tomentose, the
                    hairs not distinctly concealing the blade surface
                        9. Branchlets of the season angled, distinctly and sometimes densely hairy;
                        lower surface of leaf distinctly hairy
                            10. Hairs of the branchlets mostly erect..........*Vitis cinerea* var. *cinerea*
                            10. Hairs of the branchlets mostly appressed...............................
                            ....*Vitis cinerea* var. *floridana*
                        9. Branchlets of the season round, glabrous or only sparsely hairy; lower
                        surface of leaf glabrous or sparsely hairy, the hairs mostly (but not
                        completely) confined to the veins and vein axils
                            11. Mature leaves usually 3-lobed, sometimes deeply so, the terminal
                            lobe distinctly long-acuminate; branchlets of the season purplish red
                            throughout..*Vitis palmata*
                            11. Mature leaves usually not lobed (generally similar in form to those
                            of *V. rotundifolia*); branchlets of the season green, brown, or gray,
                            purplish only along one side..............................*Vitis vulpina*

Ampelopsis arborea **(Linnaeus) Koehne.** Peppervine. Hammocks, upland woods, flood-plains, swamp margins. Essentially throughout except the Keys.
Ampelopsis cordata **Michaux.** Raccoon-Grape, Heartleaf Peppervine. Wet woods, stream banks, floodplains. Central panhandle, Jackson, Gadsden, Liberty, and Leon counties.
Cissus trifoliata **(Linnaeus) Linnaeus.** Sorrel Vine, Marine Vine, Marine Ivy. Dunes, shell mounds, coastal hammocks. Scattered in the panhandle, more common along the

coasts of the south-central and southern peninsula; Hernando and Flagler counties southward, including the Keys.

Cissus verticillata **(Linnaeus) Nicolson & C. E. Jarvis.** Possum Grape, Season Vine. Hammocks, low woods. Southern peninsula, Brevard, Lee, and Hendry counties southward, including the Keys.

Parthenocissus quinquefolia **(Linnaeus) Planchon.** Virginia Creeper, Woodbine. Upland and lowland woods, disturbed sites. Essentially throughout, including the Keys.

Vitis aestivalis **Michaux.** Summer Grape. Upland and lowland woods, scrub, dunes, hammocks. Essentially throughout, except the Keys.

Vitis cinerea **(Engelmann) Engelmann ex Millardet var.** *cinerea.* Downy Grape, Sweet Winter Grape. Floodplains, low woods, pond and stream margins. Central panhandle, Jackson, Wakulla, and Washington counties.

Vitis cinerea **(Engelmann) Engelmann ex Millardet var.** *floridana* **Munson.** Simpson's Grape, Florida Grape. Floodplains, low woods, pond and stream margins. Essentially throughout, Okaloosa to Nassau counties, south to Monroe and Miami-Dade counties; absent from the Keys.

Vitis palmata **Vahl.** Catbird Grape, Cat Grape, Red Grape. Floodplains of large rivers. Panhandle, Santa Rosa, Jackson, Liberty, Gadsden, and Suwannee counties.

Vitis rotundifolia **Michaux.** Muscadine, Bullace. Upland and lowland woods, hammocks, floodplains, disturbed sites, woodland margins. Essentially throughout, including the Keys.

Vitis shuttleworthii **House.** Calusa Grape, Calloose Grape. Hammocks, low woods, mixed uplands. Peninsula, Citrus, Marion, and Flagler counties southward to Miami-Dade County; absent from the Keys.

Vitis vulpina **Linnaeus.** Frost Grape, Winter Grape, Chicken Grape. Mixed upland woods, wet and moist hammocks. Panhandle and northern peninsula, Escambia to Putnam and Flagler counties, south to Manatee County.

Ximeniaceae (Ximenia family)

Ximenia americana **Linnaeus.** Hog Plum, Tallow Wood. Hammocks, scrub, pinelands, swamps. Peninsula, Levy, Alachua, and Duval counties southward, including the Keys.

Zamiaceae (Zamia Family)

1. Petiole armed with prickles..*Zamia furfuracea*
1. Petiole lacking prickles...*Zamia pumila*

Zamia furfuracea. Cardboard Coontie, Cardboard Cycad, Cardboard Palm. Disturbed sites, escaped from cultivation. Southeast peninsula, Martin to Miami-Dade counties.
Zamia pumila **Linnaeus.** Coontie, Florida Arrowroot. Peninsula, from about Taylor and Suwannee counties southward, including the Keys.

Zygophyllaceae (Caltrop Family)

1. Leaflets usually and predominantly 4; corolla finely hairy; fruit 2-winged.................
....*Guaiacum officinale*
1. Leaflets usually and predominantly 6–10; corolla glabrous; fruit 5-winged................
....*Guaiacum sanctum*

Guaiacum officinale **Linnaeus.** Common Lignum Vitae. Hammocks, escaped from cultivation. Miami-Dade County. Tropical America.
!*Guaiacum sanctum* **Linnaeus.** Lignum Vitae, Holywood, Tree of Life. Subtropical hammocks. Miami-Dade County and the Keys. Cultivated beyond its native range. State endangered.

Index to Genera

Index to Common Names